T0318751

Machine Learning for Transportation Research and Applications

Machine Learning for Transportation Research and Applications

Yinhai Wang
Smart Transportation Applications and Research Laboratory
Department of Civil and Environmental Engineering
University of Washington
Seattle, WA, United States

Zhiyong Cui
School of Transportation Science and Engineering
Beihang University
Beijing, China

Ruimin Ke
Department of Civil Engineering
University of Texas, El Paso
El Paso, TX, United States

ELSEVIER

Elsevier
Radarweg 29, PO Box 211, 1000 AE Amsterdam, Netherlands
The Boulevard, Langford Lane, Kidlington, Oxford OX5 1GB, United Kingdom
50 Hampshire Street, 5th Floor, Cambridge, MA 02139, United States

Notices

Knowledge and best practice in this field are constantly changing. As new research and experience broaden our understanding, changes in research methods, professional practices, or medical treatment may become necessary.

Practitioners and researchers must always rely on their own experience and knowledge in evaluating and using any information, methods, compounds, or experiments described herein. In using such information or methods they should be mindful of their own safety and the safety of others, including parties for whom they have a professional responsibility.

To the fullest extent of the law, neither the Publisher nor the authors, contributors, or editors, assume any liability for any injury and/or damage to persons or property as a matter of products liability, negligence or otherwise, or from any use or operation of any methods, products, instructions, or ideas contained in the material herein.

ISBN: 978-0-323-96126-4

For information on all Elsevier publications
visit our website at https://www.elsevier.com/books-and-journals

Publisher: Joseph P. Hayton
Acquisitions Editor: Kathryn Eryilmaz
Editorial Project Manager: Aleksandra Packowska
Production Project Manager: Kiruthika Govindaraju
  & Selvaraj Raviraj
Cover Designer: Mark Rogers

Typeset by VTeX

Working together
to grow libraries in
developing countries

www.elsevier.com • www.bookaid.org

Contents

Companion Web Site:
https://www.elsevier.com/books-and-journals/book-companion/9780323961264

About the authors

Yinhai Wang—Ph.D., P.E., Professor, Transportation Engineering, University of Washington, USA. Dr. Yinhai Wang is a fellow of both the IEEE and American Society of Civil Engineers (ASCE). He also serves as director for Pacific Northwest Transportation Consortium (PacTrans), USDOT University Transportation Center for Federal Region 10, and the Northwestern Tribal Technical Assistance Program (NW TTAP) Center. He earned his Ph.D. in transportation engineering from the University of Tokyo (1998) and a Master in Computer Science from the UW (2002). Dr. Wang's research interests include traffic sensing, transportation data science, artificial intelligence methods and applications, edge computing, traffic operations and simulation, smart urban mobility, transportation safety, among others.

Zhiyong Cui—Ph.D., Associate Professor, School of Transportation Science and Engineering, Beihang University. Dr. Cui received the B.E. degree in software engineering from Beijing University in 2012, the M.S. degree in software engineering from Peking University in 2015, and the Ph.D. degree in civil engineering (transportation engineering) from the University of Washington in 2021. Dr. Cui's primary research focuses on intelligent transportation systems, artificial intelligence, urban computing, and connected and autonomous vehicles.

Ruimin Ke—Ph.D., Assistant Professor, Department of Civil Engineering, University of Texas at El Paso, USA. Dr. Ruimin Ke received the B.E. degree in automation from Tsinghua University in 2014, the M.S. and Ph.D. degrees in civil engineering (transportation) from the University of Washington in 2016 and 2020, respectively, and the M.S. degree in computer science from the University of Illinois Urbana–Champaign. Dr. Ke's research interests include intelligent transportation systems, autonomous driving, machine learning, computer vision, and edge computing.

Chapter 1

Introduction

This book aims to help current and future transportation professionals build their understanding of, and capability to use, machine learning (ML) methods and tools to address transportation challenges. Although this book is designed as an entry level textbook for college or graduate students, it can also serve as a reference book for working professionals on ML research and applications in transportation engineering.

This book does not require any prior experience nor knowledge of computer programming or the concepts of ML and its transportation related applications to read it. Readers can gradually build the needed programming skills through reading this book and working on the exercises in each chapter. Considering the breadth of the topics covered in this book and the frequent updates needed for the programming scripts and the supporting packages, the authors choose to provide the computer codes for all the exercises online rather than in the book. Also, to make it easier for instructors to teach courses relevant to ML in transportation using this textbook, the authors will share PowerPoint files for each chapter and solutions to the example problems in the companion website of this book.

The remaining part of this chapter introduces the general background of transportation and ML and explains why the authors consider this book to be needed for transportation education, followed by the organization of this book.

1.1 Background

1.1.1 Importance of transportation

Transportation is a means of moving goods, humans, and animals from place to place. It is essential for everyone's daily life. Every aspect of modern economies, and the ways of life they support, can be tied directly or indirectly to transportation. Thus, transportation is very important to the economy. At the microscopic level, a typical household spends about 15% of its income on transportation. Traffic congestion alone costs each American driver $1,377, indicating a total congestion cost of $88 billion in the US in 2019, according to the INRIX 2019 Global Traffic Scorecard. Transportation is the second largest household expenditure category in the US when household spending, such as healthcare benefits, is excluded. The 2020 data show that an average of $9,826, or 16%, was spent on transportation by households in the US. At the macroscopic level, many wealthy countries spend approximately 6% to 12% of the Gross Domestic Product (GDP) on transportation (Rodrigue, 2020). In the US, transportation contributed 8% to the 2020 GDP and is the fourth largest contributor following housing, healthcare, and food.

Transportation plays a critical role in climate change. Transportation consumed 28% of total energy use in the US in 2021. Petroleum is the main source of energy for transportation in the US, accounting for 90% of the total transportation sector energy use. Consequently, transportation has been the largest contributor to US greenhouse gas emissions, account for 27% of the total. The increase of greenhouse gas emissions in the transportation sector is greater than any other sectors in absolute terms between 1990 and 2020.

Transportation also impacts public health significantly. Air pollutants emitted from transportation vehicles lead to poor air quality that has negative impacts on public health. Many health problems, such as respiratory infections, cardiovascular diseases, lung cancers, etc., have been found to be related to traffic pollution. Also, traffic crashes kill approximately 1.35 million people each year globally. Road traffic collisions are the leading cause of death among people aged 5–29 according to the 2018 Global Status Report on Road Safety published by the World Health Organization. Although many new safety solutions have been applied to vehicles and roadways, traffic fatalities are still high in the US, with approximately 40,000 people killed each year. Pedestrian fatalities have also been increasing after hitting the lowest recent number in 2009.

1.1.2 Motivation

Transportation is important. Improvements in the transportation system can generate remarkables benefit for the economy and society. Investment in transportation infrastructure can generate a 5% to 20% annual return (Rodrigue, 2020). However, transportation is a very complicated system-of-systems through which humans, technologies, and infrastructure interact. Our transportation system is about how people connect, build, consume, and work, with many variables involved, including policy, human, geographical, and technical factors, among others. To gain a good understanding of the transportation issues and identify effective solutions, quality data are essential.

Over the recent decades, transportation agencies have made significant investments in system sensing and data gathering. For example, data exchange and AI/ML are both listed as fundamental elements of intelligent transportation systems (ITS), a system of technologies and operational advancements that improve the capacities of the overall transportation system. The global investment on ITS technologies is projected to grow at a compounded annual rate of 8.8%, reaching $49.5 billion by 2026. These new ITS system data, third-party data (such as those from navigation apps), and classical traffic sensor and survey data form big data streams that enable system-wide in-depth analyses concerning a variety of important transportation issues.

Unfortunately, classical transportation research and practice have been mainly based on the very limited amounts of data collected from periodical surveys or limited sensor locations. Most methods, including the commonly used ones, are based on mathematical assumptions without sufficient validation (Ma et al., 2011) and are not suitable for using such new datasets. Transportation curricula currently used in universities are strong in design, construction, and operation topics, but lack coverage of database, data analytics, and information technologies. Consequently, transportation agencies and companies are facing increasing workforce challenges, particularly for the use of new technologies such as sensing, Internet of Things (IoT), connected and automated vehicles, drones, big data analytics, and artificial intelligence (AI).

To help address this workforce challenge, the authors decided to write this book and focus on ML because it has the potential to "transform ITS at every level of implementation." (Chan-Edmiston et al., 2020) ML has been widely applied in transportation data collection, autonomous driving, transportation asset management, demand forecasting, safety improvement, mobility as a service, etc. It is clearly a critical skill for future transportation professionals. This book provides an introduction to ML methods for those transportation students and working professionals who are interested in studying this subject. Hopefully, this book lays a solid foundation for addressing the current and future workforce challenges in transportation.

1.2 ML is promising for transportation research and applications

1.2.1 A brief history of ML

Before presenting ML, we need to briefly introduce AI, a research field which started at a workshop held on the campus of Dartmouth College, NH, in the summer of 1956 (Kaplan and Haenlein, 2019). The term of AI was first coined at this workshop by John McCarthy, a professor of computer science at Stanford University. Although there is not a universally accepted definition of AI, the key concept is the same: AI leverages computers and machines to mimic the problem-solving and decision-making capabilities of the human mind, as described on the IBM website at https://www.ibm.com/cloud/learn/what-is-artificial-intelligence. Research investigation in whether machine could substitue human for some jobs can be dated back to 1945 when Vannevar Bush proposed a system to amplify people's own knowledge and understanding (Bush et al., 1945). Machine intelligence was more clearly described in 1950 by Alan Turing, one of the most influential British mathematician and logician who made major contributions to mathematics, cryptanalysis, logic, philosophy, and mathematical biology and also to the new areas later named computer science, cognitive science, artificial intelligence, and artificial life. In his paper (Turing and Haugeland, 1950), Turing proposed a method for answering the question he raised at the beginning of his paper "Can machines think?" This method, now referred to as the Turing test, can be used to test a machine's ability to exhibit intelligent behavior equivalent to, or indistinguishable from, that of a human.

Over recent decades, AI has attracted a substantial amount of attention and has been intensively studied. With the ubiquitous use of smart phones and deployment of Internet of Things (IoT), data has become increasingly available that has further stimulated AI research, particularly ML. Although AI and ML are often used exchangeably, these two terms are actually two different concepts even though ML is actually a part of AI. ML is an application of AI that focuses on the use of data and algorithms to imitate the way that humans learn, gradually improving its accuracy. ML algorithms build a model based on sample data, known as training data, in order to make predictions or decisions without being explicitly programmed to do so.

ML originated from the mathematical modeling of neural networks in 1943. Walter Pitts and Warren McCulloch (McCulloch and Pitts, 1943) attempted to mathematically map out thought processes and decision making in human cognition. Because of the emerging data sets and new computing technologies, ML technologies have advanced very quickly, making them significantly more effective than in the past. A remarkable milestone is the development of AlphaGo, the first computer program defeated in 2016 a Go world champion, Lee Sedol, and is arguably the strongest Go player in history. AlphaGo is a computer program developed by Google DeepMind to play the board game Go. AlphaGo's algorithm uses a combination of machine learning and tree search techniques, combined with extensive training, both from human and computer play. AlphaGo was trained on thousands of human amateur and professional games to learn how to play Go. Its improved version, AlphaGo Zero, released just roughly a year after, skips this human Go game data based training step and learns to play simply by playing games against itself, starting from completely random play. In doing so, it quickly surpassed human level of play and defeated the previously published champion-defeating version of AlphaGo by 100 games to 0. This example demonstrates the power of ML.

1.2.2 ML for transportation research and applications

Although the history of using ML in transportation is short, researchers have made remarkable progress demonstrating the high value of the method. From transportation data collection to transportation operations, demand forecasting, and planning, ML methods have been proven effective over the conventional methods. Here are some examples.

- Traffic data collection. This is the first and foremost step for ITS.It is the foundation for traffic system control, demand prediction, infrastructure monitoring and management, etc. It is also the most important technology for autonomous vehicles. Traffic data collection, or traffic sensing, from the functionality perspective, has various manners including infrastructure-based sensing, vehicle onboard sensing, and aerial sensing. Due to the different characteristics of traffic sensing technologies, how to extract useful traffic data and merge different types of sensing technologies together is critical for the future transportation system. ML methods have been used for vehicle detection and classification, road surface condition sensing, pedestrian and bicyclist counting, etc.

- Traffic prediction. Coupled with traffic sensing, traffic state prediction is another motivating application. The traffic state variables include traffic speed, volume, travel time, and other travel demand factors, such as origin and destinations (OD). The road traffic state prediction considered as a time series forecasting problem can be easily solved by various types methods. However, when comes to the traffic prediction for large-scale city areas with hundreds or thousands of road segments involved, the common time series forecasting methods may not work. Thus, how to extract comprehensive spatial-temporal features from a road network to fulfill traffic prediction and assist further traffic management is a key task for transportation planning and operations. ML methods have been widely used for short and long term traffic forecasts.

- Traffic system operations. AI traffic management is poised to revamp urban transportation, relieving bottlenecks and choke-points that routinely snarl our urban traffic. This helps reduce not only congestion and travel time but also vehicle emissions. Traffic congestion mostly occurs due to the negligence of certain factors like distance maintained between two moving vehicles, traffic lights, and road signs. Congestion leads to higher fuel consumption, increased air pollution, unnecessarily wastage of time & energy, chronic stress & other physiological problems, whereas higher traffic violations are the major cause of road fatalities. Intelligent traffic management systems refer to the usage of AI, machine learning, computer vision, sensors, and data analysis tools to collect and analyze traffic data, generate solutions, and apply them to the traffic infrastructure. AI can use live camera feeds, sensors, and even Google Maps to develop traffic management solutions that feature predictive algorithms to speed up traffic flow. Siemens Mobility recently built an ML-based monitoring system that processes video feeds from traffic cameras. It automatically detects traffic anomalies and alerts traffic management authorities. The system is effective at estimating road traffic density to modulate the traffic signals accordingly for smoother movement. ML methods will improve many smart transportation applications including emergency vehicle preemption, transit signal priority, and pedestrian safety.

1.3 Book organization

The chapters in this book are organized based on the optimal learning order of the selected machine learning methods. In each chapter, we first introduce machine learning algorithms and then present related transportation case studies into which those algorithms can be applied. Specifically, the book is organized to include the following chapters:

- Chapter 2 covers the various transportation data and sensing technologies including infrastructure-based, vehicle onboard, and aerial sensing technologies and the corresponding generated datasets. This chapter also introduces existing transportation data and sensing challenges.
- Chapter 3: This chapter introduces a spectrum of key concepts in the field of machine learning, starting with the definition and categories of machine learning, and then covering the basic building blocks of advanced machine learning algorithms. The theory behind the common regressions, including linear regression and logistic regression, gradient descent algorithms, regularization, and other key concepts of machine learning are discussed.
- Chapter 4: This chapter introduces neural network techniques starting from linear regression to the feed-forward neural network (FNN). Basic FNN components, including layers, activation functions, back-propagation algorithms, and training strategies are introduced. Additionally, case studies of representative transportation applications using FNN are presented.
- Chapter 5: This chapter introduces the fundamental mechanism of convolutional neural network (CNN) which has been widely used to learn matrix-like data, such as images. The case studies introduced in this chapter includes traffic video sensing and spatiotemporal traffic pattern learning.
- Chapter 6: This chapter introduces the fundamental mechanism of recurrent neural network (RNN) and its famous variants, like LSTM and GRU. We also present transportation related case studies adopting RNN including road traffic prediction and traffic time series data imputation.
- Chapter 7: This chapter introduces reinforcement learning (RL), which is a branch of machine learning targeting to solve these sequential decision problems. In this chapter, the basic concepts, such as Markov decision process (MDP), and value-based and policy-based algorithms of RL are introduced in detail. Multi-agent reinforcement learning (MARL) aiming at controlling a bunch of objects/agents cooperatively to achieve a better system performance is also briefly introduced in this chapter. We also present typical simulated and real scenarios to apply different RL algorithms in transportation applications, such as traffic signal control and car following problems.
- Chapter 8: This chapter introduces transfer learning and its applications in improving intelligent transportation systems, such as enhancing parking surveillance and traffic volume detection.
- Chapter 9: Graph neural network (GNN) as a building block of deep neural networks has the superiority in extracting and processing comprehensive features from graph data and enhances the interpretability of neural network models. This chapter describes the basic concepts of GNN and state-of-the-art GNN variants, such as graph convolutional neural network (GCN). In addition, a variety of transportation applications using GNN, including traffic signal control and road traffic prediction, are introduced.
- Chapter 10: This chapter introduces a representative generative model, i.e. generative adversarial network (GAN). is an unsupervised learning task that

involves automatically mining and learning the patterns or distributions in a dataset to generate new data samples with some variants. We introduce the theoretical background of generative models and the details of the frameworks of and GAN variants. GAN-related case studies in the transportation domain, such as traffic state estimation, are also briefly introduced in this chapter.

- Chapter 11: Edge computing is crucial for the future ITS applications to enable the interaction between connected traffic participants. This chapter presents representative edge computing scenarios in the ITS field and introduces the background and fundamental concepts of federated learning (FL), which is a method for distributed model training across many edge computing devices.
- Chapter 12: AI will definitely be one of the fundamentals of future transportation systems. This chapter introduces key aspects and directions of AI techniques that will hugely impact and advance the urban transportation systems in the future. Additionally, this chapter also discuss the extension and future plan of this book.

Chapter 2

Transportation data and sensing

2.1 Data explosion

Urbanization has been posing great opportunities and challenges in multiple areas, including the environment, health care, the economy, housing, transportation, among others. The opportunities and challenges boost the rapid advances in cyber-physical technologies and bring connected mobile devices to people's daily life. Nowadays, approximately 90% of people are connected to the internet and have fast access to a wide variety of information. The superb communication infrastructure in urban area has been attracting more and more residents to cities at an unprecedented scale and speed. In order to efficiently manage the data generated every day and to use them to better allocate urban resources, the smart city concept has emerged. This concept combines sensors, system engineering, artificial intelligence (AI), and information and communication technologies to optimize city services and operations. Naturally, the transportation system that moves goods and people is a critical component of the smart city. Intelligent Transportation Systems (ITS) has likewise emerged as a concept that applies sensing, analysis, control, and communications technologies to ground transportation in order to improve safety, mobility, efficiency, and sustainability.

Modern ITS applications are data-driven and make high demands for computing services and sophisticated models to process, analyze, and store big data. Transportation data, as a major data source in urban computing, is characterized by its high volume, high heterogeneity in data formats, and high variance in data quality. Machine learning has been extensively applied to ITS and greatly leverages the power and potential of ITS data. Machine learning models are often data-hungry and computationally expensive compared to traditional statistical models. The training of a machine learning model, especially deep neural networks (DNNs), may take days or weeks to complete. In most cases, making inferences from machine learning models is also less efficient than traditional methods. Understanding traffic data types and properties is a critical step for comprehending machine learning applications in ITS.

Traffic data collection, or, in other words, traffic sensing, is the first and foremost step in ITS since it is the foundation for modeling, system analysis, traffic prediction, and control. Traffic data collection and sensing, from the functionality perspective, can be divided into infrastructure-based sensing, vehicle onboard sensing, and aerial sensing. Transportation infrastructures and road users are the two major components of ground transportation systems, and aerial sensing using aircraft such as drones is an emerging technology and can be considered an extension of ground transportation systems.

2.2 ITS data needs

Data collection is the first step in most ITS applications. Advanced traffic management systems (ATMS) need traffic operation data, e.g., inductive loop data and surveillance camera data, to monitor the roadways and intersections. Advanced traveler information systems (ATIS) need traffic and environmental data and communicates with travelers regarding congestion, incident, work zone, extreme weather, etc. Advanced vehicle control systems (AVCS) need information on the subject vehicles, as well as that of the surrounding environment, to operate certain advanced vehicular functions, such as collision warning and adaptive cruise control. In Commercial Vehicle Operations (CVO), weigh-in-motion data and sensor data regarding the containers are critical in monitoring truck operations and safety. These subsystems of ITS all rely heavily on high-quality ITS data, often from multiple sources, to function in the expected manner.

ITS data are generated from sensors (Klein et al., 2006). ITS sensors have their own working mechanisms, pros/cons, and application scenarios. The inductive loop detector detects vehicles via loop inductance decreases when a vehicle is on top of it. Inductive loop detection is a well-understood technology that provides basic traffic parameters and is insensitive to inclement weather. But the loop detector's installation and maintenance require pavement cuts and lane closures. The magnetic sensor detects the earth's magnetic field change when a vehicle is over it and can be used where loops are not feasible, e.g., on bridge decks. They are less susceptible than loops to the stresses of traffic, but cannot detect motionless vehicles unless special sensor layouts and signal processing software are used. Pneumatic road-tube sensors are portable and can be set up easily on roads to count vehicles by using air pressure signals. They have a low power usage and are lower cost to maintain, yet are temperature sensitive and may produce inaccurate axle counting when truck and bus volumes are high. Microwave radar sensors can generate speed and multi-lane count data with high accuracy, but generate no vehicle classification information. Infrared sensors can detect vulnerable road users in addition to vehicles, but have reduced vehicle sensitivity in heavy rain, snow, and dense fog. Traffic cameras are a type of sensor for collecting rich traffic information and are widely deployed, but the data production process sometimes requires advanced computer vision algorithms and powerful computers. Acoustic sensors, ultrasonic sensors, LiDAR, etc.are other popular ITS sensors that support ITS data collection.

According to Ke (2020), modern ITS systems operations, from both the functionality and the methodological perspectives, can be roughly divided into surface transportation-infrastructure systems, surface transportation-vehicle systems, and aerial transportation systems as an extension. These three components each serve different purposes and pose different properties to be leveraged in data-driven application development. In the following three sections, the data and example ITS data-driven applications are introduced based this categorization.

2.3 Infrastructure-based data and sensing

A key objective of the ITS concept is to leverage the existing civil infrastructures to improve traffic performance. Transport infrastructure refers to roads, bridges, tunnels, terminals, railways, traffic controllers, traffic signs, other roadside units, etc. Sensors installed within transport infrastructures monitor certain locations in a transportation system, such as intersections, roadway segments, freeway entrances/exists, and parking facilities.

2.3.1 Traffic flow detection

One of the fundamental functions of infrastructure-based ITS sensing is traffic flow detection and classification at particular locations. Vehicle counts, flow rate, speed, density, trajectories, classes, and many other valuable data can be made available through traffic flow detection and classification. Chen et al. (2020b) proposed a traffic flow detection method using optimized YOLO (You Only Look Once) for vehicle detection and DeepSORT (Deep Simple Online and Realtime Tracking) for vehicle tracking and implemented the method on the edge device Nvidia Jetson TX2. Haferkamp et al. (2017) proposed a method by applying machine learning (KNN and SVM) to radio attenuation signals and were able to achieve success in traffic flow detection and classification. If processed with advanced signal processing methods, traditional traffic sensors, such as loop detectors and radar, can also expand their detection categories and performance. Ho and Chung (2016) applied Fast Fourier Transforms (FFTs) to radar signals to detect traffic flow at the roadside. Ke et al. (2018b) developed a method for traffic flow bottleneck detection using wavelet transforms on loop detector data. Distributed sensing with acoustic sensing, traffic flow outlier detection, deep learning, and robust traffic flow detection in congestion are examples of other state-of-the-art studies in this subfield (Liu et al., 2020c, 2018; Djenouri et al., 2018; Liang et al., 2020).

2.3.2 Travel time estimation

Coupled with traffic flow detection, travel time estimation is another task in ITS sensing. Accurate travel time estimation needs multilocation sensing and reidentification of road users. Bluetooth sensing is the primary way to detect travel time since Bluetooth detection comes with a MAC address of a device so it can naturally reidentify road users that carry the device. Vehicle travel time (Martchouk et al., 2011) and pedestrian travel time (Malinovskiy et al., 2012) can both be extracted with Bluetooth sensing, but Bluetooth sensing has generated privacy concerns. With advances in computer vision and deep learning, travel time estimation has improved for road user reidentification using surveillance cameras. Deep image features are extracted for vehicles and pedestrians and then compared employing region-wide surveillance cameras for multi-camera tracking (Liu et al., 2016; He et al., 2020; Lee et al., 2020; Han et al., 2020b,a). An effective and efficient pedestrian reidentification method was developed by Han et al. (2020b), called KISS+ (Keep It Simple and Straight-forward Plus), in which multifeature fusion and feature dimension reduction are performed based on the original KISS method. Sometimes it is not necessary to estimate travel time for every single road user. In such cases, more conventional detectors and methods could achieve good results. As Oh et al. (2002) proposed a method to estimate link travel time, using loop detectors. The key idea was based on road section density that can be acquired by ob-serving in-and-out traffic flows between two loop stations. While no reidentification was realized, these methods had reasonably good performances and provided helpful travel time information for traffic management and users (Oh et al., 2002; Stehly et al., 2007; Cortes et al., 2002).

2.3.3 Traffic anomaly detection

Another topic in infrastructure-based sensing is traffic anomaly detection. As the name suggests, traffic anomaly refers to such abnormal incidents in an ITS. They rarely occur, and examples include vehicle breakdown, collision, near-crash, wrong-way driving, among others. Two major challenges in traffic anomaly detection are (1) the lack of sufficient anomaly data for algorithm development and (2) the wide variety of anomalies that have no clear definition. Anomalies detection is achieved mainly using surveillance cameras, given the requirement for rich information, though time series data is also feasible in some relatively simple anomaly detection tasks (Mercader and Haddad, 2020). Traffic anomaly detection can be divided into three categories: supervised learning, unsupervised learning, and semi-supervised learning. Supervised learning methods are useful when the number of classes is clearly defined and training data is large enough to make the model statistically significant; but supervised learning requires manual labeling and needs both data and labor, and they cannot detect unforeseen anomalies (Singh and Mohan, 2018; Lee et al., 2018; Loewenherz et al., 2017). Unsupervised learning has no requirement for labeling data and is more generalizable to the unforeseen anomaly as long as sufficient normal data is provided; however, anomaly detection is difficult when the data nature changes over time (e.g., if a surveillance camera keeps changing its angle and direction) (Roshtkhari and Levine, 2013). Li et al. (2020c) designed an unsupervised method based on multi-granularity tracking, and their method won first prize in the 2020 AI City Challenge. Semi-supervised learning needs only weak labels. Chakraborty et al. (2018) proposed a semi-supervised model for freeway traffic trajectory classification using YOLO, SORT, and maximum-likelihood-based Contrastive Pessimistic Likelihood Estimation (CPLE). This model detects anomalies based on trajectories and improves the accuracy by 14%. Sultani et al. (2018) considered videos as bags and video segments as instances in multiple instance learning and automatically learned an anomaly ranking model with weakly labeled data. Recently, traffic anomaly detection has been advanced not only by the design of new learning methods but also by object tracking methods. It is interesting to see that, in the 2021 AI City Challenge, all top-ranking methods somewhat made contributions to the tracking part (Zhao et al., 2021; Wu et al., 2021; Chen et al., 2021).

2.3.4 Parking detection

In addition to roadway monitoring, parking facility monitoring, as another typical element in urban areas, plays a crucial role in infrastructure-based sensing. Infrastructure-based parking-space detection can be divided into two categories from the sensor functionality perspective: the wireless sensor network (WSN) solution and camera-based solution. The WSN solution has one sensor for each parking space, and the sensors need to be low power, sturdy, and affordable (Lin et al., 2017; Lou et al., 2019; Park et al., 2008; Lee et al., 2008; Zhang et al., 2013, 2014; Jeon et al., 2018; Grodi et al., 2016; Sifuentes et al., 2011; Zhu and Yu, 2015). The WSN solution has some pros and cons: algorithm-wise, it is often straightforward; while a thresholding method would work in most cases, a relatively simple detection method may lead to a high false detection rate. A unique feature of the WSN is that it is robust to sensor failure due to the large number of sensors. This means that, even if a malfunction, the WSN still covers most of the spaces. However, a large number of sensors do require high costs for labor and maintenance in large-scale installation. Magnetic nodes, infrared sensors, ultrasonic sensors, light sensors, and inductance loops are the most popular sensors. For example, Sifuentes et al. (2011) developed a cost-effective parking space-detection algorithm based on magnetic nodes that integrates a wake-up function with optical sensors. The camera-based solution has become increasingly popular due to advances in video sensing, machine learning, and data communication technologies (Lin et al., 2017; Bulan et al., 2013; Cho et al., 2018; Amato et al., 2017; Nurullayev and Lee, 2019; Alam et al., 2018; Wu et al., 2007; Rianto et al., 2018; Baroffio et al., 2015; Amato et al., 2016; Vítek and Melničuk, 2017; Ling et al., 2017; Nieto et al., 2018). Compared to the WSN, one camera covers multiple parking spaces; thus, the cost per space is reduced. It is also simpler to maintain since the installation of camera systems is nonintrusive. Additionally, as aforementioned, video contains more information than other sensors, so it has the potential for applications to more complicated parking tasks. Bulan et al. proposed to use background subtraction and SVM for street parking detection; it achieved a promising performance level and was not sensitive to occlusion (Bulan et al., 2013). Nurullayev et al. designed a pipeline with a unique design using a dilated convolutional neural-network (CNN) structure. The design was validated to be robust and suitable for parking detection (Nurullayev and Lee, 2019).

2.4 Vehicle onboard data and sensing

Vehicle onboard sensing is complementary to infrastructure-based sensing. It happens on the road user side. The sensors move with road users, and thereby are more flexible and cover larger areas. Additionally, vehicle onboard sensors are the eyes of an intelligent vehicle, making the vehicle sense and understand the surroundings. These properties pose opportunities for urban sensing and autonomous driving technologies, but, at the same time, create challenges for implementation. A major technical challenge is the irregular movement of the sensors. Traditional ITS sensing on infrastructure-mounted sensors deal with stationary backgrounds and relatively stable environmental settings. For instance, radar sensors for speed measurement know where the traffic is supposed to be. Camera sensors have a fixed video background so that traditional background modeling algorithms can be applied. Therefore, in order to benefit from vehicle onboard sensing, it is necessary to address such challenges.

2.4.1 Traffic near-crash detection

A traffic near crash or near miss is the conflict between road users that has the potential to develop into a collision. Near-crash detection using onboard sensors is the first step for multiple ITS applications: near-crash data serves as (1) surrogate safety data for a traffic safety study, (2) corner-case data for autonomous vehicle testing, and (3) input to collision avoidance systems. There were some pioneer studies on automatic near-crash data extraction on the infrastructure side using LiDAR and cameras (Ismail et al., 2009; Wu et al., 2018; Huang et al., 2020). In recent years, near-crash detection systems and algorithms using onboard sensors have developed at a rapid pace. Ke et al. (2017) and Yamamoto et al. (2020) each applied conventional machine learning models (SVM and random forest) in their near-crash detection frameworks and achieved fairly good detection accuracy and efficiency using regular computers. State-of-the-art methods tend to use deep learning for near-crash detection. The integration of CNN, LSTM, and attention mechanisms has been demonstrated to be superior in recent studies (Yamamoto et al., 2020; Ibrahim et al., 2021; Kataoka et al., 2018). Ibrahim et al. reported that a bi-directional LSTM with self-attention outperformed a single LSTM with a normal attention mechanism (Ibrahim et al., 2021). Another feature in recent studies has been the combination of onboard camera sensor input and onboard telematics input, such as vehicle speed, acceleration, and location, to either improve the near-crash detection performance or increase the output data diversity (Yamamoto et al., 2020; Taccari et al., 2018; Ke et al., 2020a). Ke et al. mainly used onboard video for near-crash detection but also collected telematics and vehicle CAN data for post analysis (Ke et al., 2020a).

2.4.2 Road user behavior sensing

Human drivers can recognize and predict other road users' behaviors, e.g., pedestrians crossing the street and vehicles changing lanes. For intelligent or autonomous vehicles, automating this kind of behavior recognition process is expected as part of the onboard sensing functions (Jayaraman et al., 2020; Brehar et al., 2021; Chen et al., 2020c; Liu et al., 2020a). Stanford University's (Liu et al., 2020a) published an article on pedestrian intent recognition using onboard videos. They built a graph CNN to exploit spatiotemporal relationships in the videos, which was able to identify the relationships between different objects. While, for now, the intent prediction focused only on crossing the street or not, the research direction is clearly promising. They also published over 900 h of onboard videos online. Another study proposed by Brehar et al. (2021) on pedestrian action recognition used an infrared camera, which compensates for regular cameras at night and on foggy days or rainy days. They built a framework composed of a pedestrian detector, an original tracking method, road segmentation, and LSTM-based action recognition. They also introduced a new dataset named CROSSIR. Moreover, vehicle behavior recognition is of equal importance for intelligent or autonomous vehicles (Lyu et al., 2020; Wang et al., 2021b; Fernando et al., 2020; Zhang et al., 2020b; Mozaffari et al., 2020). Wang et al. (2021b) recently developed a method using fuzzy inference and LSTM for the recognition of vehicles' lane changing behavior. The recognition results were used for a new intelligent path-planning method to ensure the safety of autonomous driving. The method was trained and tested by NGSIM data. Another study on vehicle trajectory prediction using onboard sensors in a connected-vehicle environment was conducted. It improved the effectiveness of the Advanced Driver Assistant System (ADAS) in cut-in scenarios by establishing a new collision warning model based on lane-changing intent recognition, LSTM for driving trajectory prediction, and oriented bounding box detection (Lyu et al., 2020). Another type of road user-related sensing is passenger sensing, while, for other purposes, e.g., transit ridership sensing uses wireless technologies (Pu et al., 2020) and car passenger occupancy detection uses thermal images for carpool enforcement (Erlik Nowruzi et al., 2019).

2.4.3 Road and lane detection

In addition to road user-related sensing tasks, road and lane detection are often performed for lane departure warning, adaptive cruise control, road condition monitoring, and autonomous driving. The state-of-the-art methods mostly apply deep learning models for onboard camera sensors, LiDAR, and depth sensors for road and lane detection (Chen et al., 2019d; Fan et al., 2020; Wang et al., 2020b; Luo et al., 2020; Farag, 2020; Almeida et al., 2020). Chen et al. (2019d) proposed a novel progressive LiDAR adaption approach-aided road detection method to adapt LiDAR point cloud to visual images. The adaption contains two modules, i.e., data space adaptation and feature space adaptation. This camera-LiDAR fusion model currently maintains its position at the top of the KITTI road-detection leaderboard. Fan et al. (2020) designed a deep learning architecture that consists of a surface normal estimator, an RGB encoder, a surface normal encoder, and a decoder with connected skip connections. It applied road detection to the RGB image and depth image and achieved state-of-the-art accuracy. Alongside road region detection, an ego-lane detection model proposed by Wang et al. outperformed other state-of-the-art models in this subfield by exploiting prior knowledge from digital maps. Specifically, they employed OpenStreetMap's road shape file to assist lane detection (Wang et al., 2020b). Multilane detection has been more challenging and rarely addressed in existing works. Still, Luo et al. (2020) were able to achieve pretty good multilane detection results by adding five constraints to the Hough transform: a length constraint, parallel constraint, distribution constraint, pair constraint, and uniform width constraint. A dynamic programming approach was applied after the Hough transform to select the final candidates.

2.4.4 Semantic segmentation

Detecting the road regions at the pixel level is a type of image segmentation focusing on the road instance. There has been a trend in onboard sensing to segment the entire video frame at pixel level into different object categories. This is called semantic segmentation and is considered a must for advanced robotics, especially autonomous driving (Siam et al., 2018; Tao et al., 2020; Erkent and Laugier, 2020; Treml et al., 2016; Hung et al., 2018; Ouali et al., 2020; Yuan et al., 2019; Mohan and Valada, 2021; Cheng et al., 2020a; Cordts et al., 2016). Compared to other tasks that can usually be fulfilled using various types of onboard sensors, semantic segmentation is realized strictly using visual data. Nvidia researchers Tao et al. (2020) proposed a hierarchical multi-scale attention mechanism for semantic segmentation based on the observation that certain failure modes in the segmentation can be resolved on a different scale. The design of their attention was hierarchical so that memory usage was four times more efficient in the training process. The proposed method ranked best on two segmentation benchmark datasets. Semantic segmentation is relatively computationally expensive; thus, working towards the goal of real-time segmentation is a challenge (Siam et al., 2018; Treml et al., 2016). Siam et al. (2018) proposed a general framework for real-time segmentation and ran 15 fps on the Nvidia Jetson TX2. Labeling at the pixel level is time-consuming and is another challenge for semantic segmentation. There are some benchmark datasets available for algorithm testing, such as Cityscapes (Cordts et al., 2016). Efficient labeling for semantic segmentation and unsupervised/semi-supervised learning for semantics segmentation are interesting topics worth exploring (Erkent and Laugier, 2020; Hung et al., 2018; Ouali et al., 2020).

2.5 Aerial sensing for ground transportation data

Aerial sensing using drones, i.e., unmanned aerial vehicles (UAVs), has been performed in the military for years and recently has become increasingly tested in civil applications, such as agriculture, transportation, goods delivery, and security. Automation and the smartness of surface traffic cannot be fulfilled using only groundbased sensors. UAV extends the functionality of existing ground transportation systems with its high mobility, top-view perspective, wide view range, and autonomous operation (Menouar et al., 2017). UAV's role is envisaged in many ITS scenarios, such as flying accident-report agents (Ardestani et al., 2016), traffic enforcement (Constantinescu and Nedelcut, 2011), traffic monitoring (Huang et al., 2021), and vehicle navigation (Shao et al., 2019) and vehicle reidentification (Liu et al., 2022). While there are regulations to be established and practical challenges to be addressed, such as safety concerns, privacy issues, and short battery life problem, UAV's applications in ITS are envisioned to be another necessary step forward for transportation network automation (Menouar et al., 2017). On the road user side, UAV extends the functionality of ground transportation systems by detecting vehicles, pedestrians, and cyclists from the top view, which has a wider view range and better view angle (no occlusion) than surveillance cameras and onboard cameras. UAVs can also detects road users' interactions and traffic theory-based parameters, thereby supporting applications in traffic management and user experience improvement.

2.5.1 Road user detection and tracking

Road user detection and tracking are the initialization processes for traffic interaction detection, pattern recognition, and traffic parameter estimation. Conventional UAV-based road user detection often uses background subtraction and handcrafted features, assuming the UAV is not moving, or stitching frames in the first step (Teutsch and Krüger, 2012; Rodríguez-Canosa et al., 2012; Gomaa et al., 2018; Tsao et al., 2018; Cao et al., 2011; Breckon et al., 2009). Recent studies have tended to develop deep learning detectors for UAV surveillance (Khan et al., 2018; Carletti et al., 2018; Barmpounakis et al., 2019; Najiya and Archana, 2018; Zhu et al., 2018). Road user detection itself can acquire traffic flow parameters, such as density and counts, without any need for motion estimation or vehicle tracking. Zhu et al. (2018) proposed an enhanced Single Shot Multibox Detector (SSD) for vehicle detection with manually annotated data, resulting in high detection accuracy and a new dataset. Wang et al. (2020a) identified the challenge in UAV-based pedestrian detection, particularly at night, and proposed an image enhancement method and a CNN for pedestrian detection after dark. In order to conduct more advanced tasks in UAV sensing on the road user side, road user tracking is a must because it merges individual detection results. Efforts have been made regarding UAV-based vehicle tracking and motion analysis (Teutsch and Krüger, 2012; Rodríguez-Canosa et al., 2012; Gomaa et al., 2018; Li et al., 2016; Cao et al., 2014; Khan et al., 2017). In many previous research, existing tracking methods, such as particle filter and SORT, were directly applied and produced fairly good tracking performances. Recently, Ke et al. (2020b) developed an innovative tracking algorithm that incorporated lane detection information and improved tracking accuracy.

2.5.2 Advanced aerial sensing

Road user detection and tracking support advanced aerial ITS sensing applications. For example, in Khan et al. (2018); Kaufmann et al. (2018), the researchers developed new methods for traffic shock-wave identification and synchronized traffic flow pattern recognition under oversaturated traffic conditions. Chen et al. (2020a) conducted a thorough study on traffic conflict based on extracted road-user trajectories from UAVs. The Safety Space Boundary concept in the paper is an informative design for conflict analysis. One of the most useful applications using UAV is traffic flow-parameter estimation: Traditional research in this field focused on using static UAV videos for macroscopic parameters extraction. McCord et al. (2003) led a pioneering research work to extract a variety of critical macroscopic traffic parameters, such as annual average daily traffic (AADT). Later on, a new method was proposed by Shastry and Schowengerdt (2005), in which they adopted image registration and motion information to stitch images and obtain fundamental traffic-flow parameters. Recently, Ke et al. developed a series of efficient and robust frameworks to estimate aggregated traffic-flow parameters (speed, density, and volume) (Ke, 2016; Ke et al., 2016, 2018a). Because of the potential benefits of higher-resolution data in ITS, microscopic traffic parameter estimation has been conducted (Chen et al., 2020d). Barmpounakis et al. proposed a method to extract naturalistic trajectory data from UAV videos at relatively less congested intersections using static UAV video (Barmpounakis et al., 2019). Ke et al. (2020b) developed an advanced framework composed of lane detection, vehicle detection, vehicle tracking, and traffic parameter estimation that can estimate ten different macroscopic and microscopic parameters from a moving UAV.

2.5.3 UAV for infrastructure data collection

On the infrastructure side, UAV has been utilized for some ITS sensing services, such as road detection, lane detection, and infrastructure inspection. UAVs are extremely helpful at locations that are hard for humans to reach. Road detection is used to localize the regions where traffic is detected from UAV sensing data. It is crucial to support applications such as navigation and task scheduling (Karaduman et al., 2019; Li et al., 2020b; Lin and Saripalli, 2012). For instance, Zhou et al. (2014, 2016) designed two of the popular methods for road detection in UAV imagery. While there were some studies before these two, the paper by Zhou et al. (2014) was the first targeting speeding up the road localization part using a proposed tracking method. Zhou et al. (2016) presented a fully automatic approach that detects roads from a single UAV image with two major components: road/non-road seeds generation and seeded road segmentation. Their methods were tested on challenging scenes. UAV has been intensely used for infrastructure inspection, particularly bridge inspection (Chen et al., 2019b; Jung et al., 2019; Bolourian et al., 2017; Lei et al., 2018) and road surface inspection (Biçici and Zeybek, 2021; Leonardi et al., 2018; Fan et al., 2019). Manual inspection of bridges and road surfaces is costly in terms of both time and labor. Bolourian et al. (2017) proposed an aerial framework for bridge inspection using LiDAR-equipped UAVs. It included planning a collision-free optimized path and a data-analysis framework for point cloud processing. Biçici and Zeybek (2021) developed an approach with verticality features, DBSCAN clustering, and robust plane fitting to process point clouds for automated extraction of road surface distress.

2.6 ITS data quality control and fusion

Quality data is the foundation of ITS. One of the current challenges of ITS data processing and storage is that vast quantities of data are now available from a variety of data sources and with varying data quality. High-quality data is also a requirement for machine learning. Low-quality data can lead to misclassification and inaccurate prediction/regression in a given task. Data quality control for single-source data and data fusion for multisource data are necessary as a procedure immediately after data collection in ITS applications.

Noisy data and missing data are two phenomena in ITS data-quality control. While data denoising can be done with satisfactory performance using traditional methods, such as a wavelet filter, moving average model, and Butterworth filter (Chen et al., 2020e), missing data imputation is much harder since it needs to recover the original information precisely. Another commonly applied traffic data-denoising task is trajectory data-map matching. The most popular models for this task that denoises the map matching errors are often based on the hidden Markov model. Recently, there have been many attempts to develop deep learning-based, missing-data imputation. These methods often focus on learning spatiotemporal features using deep learning models so that they are able to infer the missing values using the existing values. Given the spatiotemporal property of traffic data, the Convolutional Neural Network (CNN) is a natural choice due to its ability to learn image-like patches. We previously designed a CNN-based method for loop traffic flow-data imputation and demonstrated its improved performance over the state-of-the-art methods (Zhuang et al., 2019). Generative adversarial network (GAN) is another deep learning method that is appropriate for traffic data imputation, given its recent advances in image-like data generation. Chen et al. proposed a GAN algorithm to generate time-dependent traffic flow data. This method made two modifications to the standard GAN on using real data and introduced a representation loss (Chen et al., 2019c).

GAN has also been experimented with for travel-time imputation using probe vehicle-trajectory data. Zhang et al. developed a travel time-imputation GAN (TTI-GAN) considering network-wide spatiotemporal correlations (Zhang et al., 2021).

Multisource data fusion is critical in leveraging the power of data collected from different sources, e.g., loops, cameras, and radar. These data each come in their own formats and resolutions. From a machine learning perspective, a unified data space needs to be constructed for feature learning. In ITS, a natural way to organize data into the same space is building a map-based data dashboard as an analytical and visualization platform. This kind of platform is capable of accepting, archiving, and quality-checking traffic-sensor data from all defined regions. It incorporates data from a range of sources, such as DOT data, vendor data, and public data. DRIVE Net (Wang et al., 2016) is an example of an online system for ITS data fusion, analytics, and visualization. It can be used to quickly and automatically produce statistics for Washington State DOT's reporting needs, such as congestion evaluation, travel-time analysis, and throughput productivity evaluation. It integrates a wide range of data into this online platform through its database design, data-panel design, and functional design. Besides traditional highway and city traffic data, this system integrates multi-model data including car sharing, ferry, public transit, park and ride, elevation, and pedestrian volume. Data fusion techniques in ITS can also achieve the merging of data from different individual sensors and systems at the place of data collection. Multisource data fusion in real-time enables advanced sensing applications, e.g., camera and LiDAR data fusion in support of automated driving. The communication methods and fusion techniques need to be identified. In vehicles, the controller area network is an existing and promising communication media for multisystem integration. Early fusion of the raw sensor signals and late fusion of the sensing results are possible approaches to system-and-sensor fusion.

2.7 Transportation data and sensing challenges

2.7.1 Heterogeneity

Developing advanced ITS applications requires the adoption of different sensors and sensing methods. On a large scale, heterogeneity resides in many aspects, e.g., hardware, software, power supply, and data. Sensor hardware incorporates has a wide variety of different ITS tasks. Magnetic sensors, radar sensors, infrared sensors, LiDAR, cameras, etc., are common sensor types that each poses unique advantage in particular scenarios. These sensors vary regarding cost, size, material, reliability, working environment, sensing capability, etc. Not only is there a large variety of sensors themselves, the hardware supporting the sensing functions for storage and protection is also diverse. Also, the associated hardware may limit the applicability of sensors. A sensor with local storage is able to store data onsite for later use; a sensor with a waterproof shell is able to work outdoors, while those without one may solely be available for indoor monitoring.

Even within the same type of sensors, there can be a significant variance with respect to detailed configurations that will influence the effectiveness and applicability of the sensors. Cameras with different resolution is an example: Those with high resolution are suitable for some tasks that are not possible for low-resolution cameras, such as small object detection, while low-resolution cameras may support less complicated tasks and be more efficient and less expensive. The installation locations of the same sensors also vary. As aforementioned, sensors onboard a vehicle or carried by a pedestrian have different functions than those installed on infrastructures. Some sensors can only be installed on the infrastructures, and some are appropriate for onboard sensing. For example, loop detectors and magnetic nodes are most often on or underneath the road surface, while sensors for collision avoidance need to be onboard cars, buses, or trucks.

Software is another aspect that poses heterogeneity in ITS sensing. There is open-source software and proprietary software. Open-source software is free and flexible and can be customized for specific tasks; however, there is a relatively high risk that some open-source software is unreliable and may work only in specific settings. There are many open-source codes on platforms such as GitHub. A good open-source tool can have a massive influence on the research community, such as open codes for Mask R-CNN (He et al., 2017), which has been widely applied for traffic object-instance segmentation. Proprietary software is generally more reliable, and some software comes with customer services from the company that developed the software. These software tools are usually not free and have less flexibility for customization, and it is also difficult to know the internal sensing algorithms or design. When an ITS system is composed of multiple software tools, which is likely the case most of the time, and these tools lack transparency or flexibility regarding communication, there will be hurdles in developing efficient and advanced ITS applications.

Heterogeneous settings in ITS sensing inevitably collect a heterogeneous mix of data, such as vehicle dynamics, traffic flow, driver behavior, safety measurements, and environmental features. There are uncertainties, noises, and missing patches in ITS data. Modern ITS applications require data to be of high quality, integrated, and sometimes in real-time. Despite the improvement of sensing functionality for individual sensors at a single location, new challenges arise in the integration of heterogeneous data. New technologies also pose challenges in data collection because some data under traditional settings will be redundant, and, at the same time, new data will be required for some tasks, e.g., CAV safety and mixed autonomy.

2.7.2 High probability of sensor failure

Large-scale and real-time sensing requirements will have little tolerance for sensor failure because it may cause severe problems for the operation and safety of

ITS. A representative example is sensor failure in an autonomous vehicle, which could lead to property damage, injuries, and even fatalities. When ITS becomes more advanced, where one functional system will likely consist of multiple co-ordinated modules, the failure of one sensor could cause the malfunction of the entire system. For instance, a connected vehicle and infrastructure system may stop working because some infrastructure-mounted sensors produce no readings, resulting in interruption of the data flow for the entire system.

Sensor failure may rarely occur for every individual sensor, but, according to probability theory, if the failure probability of one sensor is p during a specific period, the failure probability among N sensors will be $1 - (1 - p)^N$. When N is large enough, the probability of sensor failure will be very high. This phenomenon is similar to the fault tolerance in cloud computing, which necessitates designing a blueprint for continuing the cloud computing ongoing work when a few machines are down. However, in ITS sensing, sensor fault tolerance is more challenging due to: (1) the hardware, software, and data heterogeneity mentioned in the last subsection; (2) the fact that, unlike cloud computing settings, sensors may be connected to different networks or even not connected to any network; (3) the potential cost and loss from a sensor fault, which could be much more serious than one in cloud computing.

This problem naturally exists and is hard to eliminate because, even when the failure probability for a single sensor is extremely low, in city-scale ITS sensing applications, where there are hundreds and even thousands of heterogeneous sensors, the probability increases significantly. Furthermore, it is not realistic to reduce the failure probability of a single sensor or device to zero in the real world.

2.7.3 Sensing in extreme cases

ITS sensing tasks that seem simple could become extraordinarily complicated or unreliable in extreme cases, such as during adverse weather, due to occlusion, and at night. A typical example is video sensing, which is sensitive to lighting changes, shadows, reflection, and so forth. In smart parking surveillance, a recent study showed that video-based detectors performed more reliably indoors than outdoors due to extreme lighting conditions and adverse weather (Ke et al., 2020d). Due to low-lighting conditions, even the cutting-edge video-based ADAS products on the market are not recommended for operation at night (Spears et al., 2017). The LiDAR sensor is one of the most reliable sensors for ITS, however, LiDAR sensing performance downgrades in rainy and snowy weather, and it is also sensitive to objects with reflective surfaces. GPS sensors experience signal obstruction due to surrounding buildings, trees, tunnels, mountains, and even human bodies. Therefore, GPS sensors work well in open areas but not in areas where obstructions are unavoidable, such as in central cities.

In ITS, especially in automated vehicle testing, extreme cases can also refer to corner cases that an automated and intelligent vehicle has not encountered before. For example, a pedestrian crossing the freeway at night may not be a common case that is thoroughly covered in the database, so a vehicle might not understand the sensing results enough to proceed confidently; therefore, it would cause uncertainty in the real-time decision-making. Some corner cases may be created by attackers. Adding noise that is unnoticeable to human eyes to a traffic sign image could result in a missed detection of the sign (Wang et al., 2021a); these adversarial examples threaten the security and robustness of ITS sensing. Corner case detection appears to be one of the hurdles that slow down the pace to achieving L-5 autonomous driving. The first question is: How does a vehicle know when it encounters a corner case? The second question is: How should it handle the unforeseen situation? We expect that corner case handling will not only be an issue for the automated vehicle but also faced for the broad range of ITS sensing components.

2.7.4   Privacy protection

Privacy protection is another major challenge. As ITS sensing becomes advanced, more and more detailed information is available, and there have been increasing concerns regarding the use of the data and the possible invasion of privacy. Bluetooth sensing detects the MAC address of the devices such as cell phones and tracks the devices in some applications, which not only risk people's identification but also their location information. Camera images, when not properly protected, may contain private information, such as faces and license plates. These data are often stored on the cloud and not owned by the people whose private information is there.

2.8   Exercises

- Name two fixed traffic sensors and two mobile traffic sensors, and their pros and cons.
- List three impacts that urban data explosion has on transportation systems.
- What are the three fundamental types of data for macroscopic traffic flow?
- What is the major difference between infrastructure-based data collection and vehicle-based data collection?
- What are the advantages and challenges of collecting surface-traffic data from aerial vehicles?
- How is travel time data collection done? Describe three methods and the core differences from other traffic data collection.
- Traffic anomaly is not well defined. What is your definition of traffic anomaly? List five traffic anomaly events. What are your recommendations for sensors and methods to collect these anomaly data?
- How could road user-behavior data help improve autonomous driving functions?
- List five traffic operation tasks that could benefit from road and lane detection.

- Briefly state the three aspects of transportation data heterogeneity and how they create a unique heterogeneous data problem in transportation.
- What is your suggestion for handling privacy issues in transportation data collection without losing the benefits that the data bring to us?
- From a probabilistic perspective, why is large-scale sensing challenging? Which transportation applications are the most sensitive to sensor failure?

Chapter 3

Machine learning basics

Machine learning has changed many industries and research domains, including urban science and transportation engineering. Machine learning methods and techniques can be categorized from different perspectives. The most common one is whether a machine learning method is supervised or not. The key difference is the presence of labels in supervised learning and the absence of labels in unsupervised learning. A label is a gold standard for a target of interest, such as the location of a vehicle in an image for training a vehicle detection model. Supervised learning has been widely used for forecasting the state of traffic network, detecting traffic objects, classifying traffic modes, etc. Unsupervised learning also has been widely applied to cluster traffic data into different groups and identify the main characteristics of those groups. Another type of machine learning method, i.e. reinforcement learning, which cannot be classified into these two categories, will be introduced in later chapters.

In this chapter, we will introduce the basic machine learning concepts and methods. We will briefly introduce several criteria for categorizing machine learning methods. Then, basic supervised learning methods, which are the basis of deep learning, and several representative unsupervised learning methods are introduced.

3.1 Categories of machine learning

There are various ways to classify machine learning problems. Before introducing specific machine learning methods, we first introduce the several categories of machine learning.

3.1.1 Supervised vs. unsupervised learning

Supervised learning is a machine learning approach that's defined by its use of labeled datasets. The datasets are designed to train or "supervise" algorithms into classifying data or predicting outcomes accurately. Using labeled inputs and outputs, the model can measure its own accuracy and learn over time. Supervised learning can be generally separated into two types of problems:

- **Classification** algorithms target accurately assigning test data into specific categories, such as separating apples from oranges. In the real world, supervised learning algorithms can be used to classify ramp-metering traffic lights into two different states, i.e., red and green.

- **Regression** is another type of supervised learning method that uses an algorithm to understand the relationship between dependent and independent variables. Regression models are helpful for predicting numerical values based on different data points, such as the traffic speed on a specific road segment. Some popular regression algorithms are linear regression and logistic regression.

Unsupervised learning uses machine learning algorithms to analyze and cluster unlabeled data sets. These algorithms discover hidden patterns in data without the need for human intervention (hence, they are "unsupervised"). Unsupervised learning models are usually used for three main tasks, clustering, association and dimensionality reduction, which can be defined as follows:

- **Clustering** is a technique for grouping unlabeled data based on their similarities or differences. For example, K-means clustering algorithms assign similar data points into groups, where the K value represents the size of the grouping and granularity. This technique is helpful for travel behavior categorization, etc.
- **Dimensionality reduction** is a learning technique used when the number of features (or dimensions) in a given dataset is too high. It reduces the number of data inputs to a manageable size, while also preserving the data integrity. Often, this technique is used in the preprocessing data stage, such as when autoencoders remove noise from visual data to improve picture quality.

There is an obvious gap between supervised learning and unsupervised learning which can be filled by semi-supervised learning. **Semi-supervised learning** is similar to supervised learning, but instead uses both labeled and unlabeled data. Labeled data is essentially information that has meaningful tags so that the algorithm can understand the data, while unlabeled data lacks that information. By using this combination, machine learning algorithms can learn to label unlabeled data.

3.1.2 Generative vs. discriminative algorithms

Generative algorithms try to model "how to populate the dataset". This reduces to learning 'everything' about the distribution that is most likely to produce what we have observed. In the probabilistic form, generative algorithms learn joint probability: $P(X, Y)$ that can be converted in Bayesian fashion: $P(X|Y) = P(Y|X)P(X)$. Some popular generative algorithms are: the naive Bayes classifier, generative adversarial networks, Gaussian mixture model, hidden Markov model, etc.

Discriminative algorithms focus on modeling a direct solution that learns the boundary instead of the distribution. For example, the logistic regression algorithm models a decision boundary. Discriminative algorithms estimate posterior probabilities. Unlike the generative algorithms, they don't model the underlying probability distributions. In the probabilistic form, the conditional probability of target Y is modeled given an observation x, $P(Y|X = x)$. Some popular discriminative algorithms are: k-nearest neighbors (k-NN), logistic regression, Support Vector Machines (SVMs), decision trees, the random forest, etc.

3.1.3 Parametric vs. nonparametric modeling

A model parameter is a configuration variable that is internal to the model and whose value can be estimated from the given data. Parameters are required by the model when making predictions. They are estimated or learned from historical training data instead of set manually by the practitioner. The learned parameters are often saved as part of the learned model.

A learning model summarizes the data with a set of fixed-size parameters (independent on the number of instances of training). **Parametric** machine learning algorithms optimize the function to a known form, while **nonparametric** machine learning algorithms are those that do not make specific assumptions about the type of the mapping function. Rather, they are prepared to choose any functional form from the training data by not making assumptions.

The word nonparametric does not mean that the value lacks existing parameters, but rather that the parameters are adjustable and can change. A simple-to-understand nonparametric model is the k-nearest neighbors' algorithm, which predicts a new data instance based on the most similar training patterns k. The only assumption it makes about the data set is that the training patterns that are the most similar are most likely to have a similar result.

A generative model is mostly parametric since it models the underlying distribution, while a discriminative model is mostly nonparametric. Note that, by definition, a model with unfixed amount of parameters is defined as nonparametric. Example models include:

- Generative/nonparametric: Gaussian mixture models (GMM) which learn a Gaussian distribution and have unfixed amount of parameters (latent parameters increases depending on the sample size);
- Generative/parametric: various Bayes-based models;
- Discriminative/parametric: Latent Dirichlet Allocation (LDA) and logistic regression;
- Discriminative/nonparametric: KNN, SVM, and K-Means.

Choosing the right approach for your situation depends on how we assess the structure and volume of the dataset, as well as the use case. To make the right decision, we need to carefully evaluate the input data with or without labels, define the goals, and review the options for algorithms. In the following sections, we will introduce several representative supervised learning and unsupervised learning algorithms.

3.2 Supervised learning

3.2.1 Linear regression

Linear regression is perhaps the most basic supervised learning algorithm. In regression, we are interested in predicting a scalar-valued target, such as the speed on a road segment. By linear, we mean that the target must be predicted as a linear function of the inputs.

Problem setup

In order to formulate a learning problem mathematically, we need to define two things: a **model** and a **loss function**.

The model, or architecture defines the set of allowable hypotheses, or functions that compute predictions from the inputs. The hypotheses contains all the possible functions $\mathcal{F} = \{f \mid y = f(x)\}$, where x and y are the variables defined in the input space $\mathcal{X}$ and $\mathcal{Y}$. In the case of linear regression, the model simply consists of linear functions. Recall that a simplest linear function, where the input and output data are one-dimensional, can be mathematically written as:

$$y = wx + b, \tag{3.1}$$

where we call the coefficient w weight, and call the intercept term bias. If the input data is multidimensional with several variables, the model can be written as

$$y = \sum_j w_j x_j + b. \tag{3.2}$$

In order to quantify how good the fit/estimation/prediction is, we define a loss function, $\mathcal{L}(\hat{y}, y)$, which measures how far off the prediction is from the target. Given a dataset $D = \{(x^{(i)}, y^{(i)})\}$ with N samples (pairs of x and y), $y^{(i)}$ is the targeted label value, and we use $\hat{y}^{(i)}$ to demonstrate the predicted value of $y^{(i)}$. In linear regression, we use squared error in the loss function, defined as

$$\mathcal{L}(\hat{y}, y) = \frac{1}{2} \sum_i (\hat{y}^{(i)} - y^{(i)})^2. \tag{3.3}$$

The loss is small when $\hat{y}$ and y are close together and large when they are far apart. In general, the value $\hat{y} - y$ is called **residual**. If the residuals is close to zero, it shows the regression model has a good fit. Note that the factor $\frac{1}{2}$ in front just makes the calculations convenient when it needs to take the derivative of the loss function.

Combining the model and the loss function, we can consider the regression needed to solve an optimization problem, where we are trying to minimize a cost function, i.e., the loss function, with respect to the model parameters (i.e., the weights w and bias b).

Note that there is a concept, cost function, that is close to the loss function. Taking the simplest linear regression as an example, the difference can be reflected by the following equation of the cost function:

$$J(w, b) = \frac{1}{2} \sum_i (y^{(i)} - wx^{(i)} - b)^2. \tag{3.4}$$

The loss is a function of the predictions and targets, while the cost is a function of the model parameters.

Solving the optimization problem

Whenever we want to solve an optimization problem, a good place to start is to compute the partial derivatives of the cost function. Just to quickly recap, given a dataset $D = \{(x^{(i)}, y^{(i)})\}$ where $x^{(i)} = \{x_1^{(i)}, \ldots, x_j^{(i)}, \ldots, x_M^{(i)}\}$, suppose f is a function of $x^{(i)}$. The partial derivative $\partial f/\partial x_j$ states in what way the value of f changes if you increase x_j by a small amount, while holding the rest of the arguments fixed. Let's do that in the case of linear regression. Applying the chain rule, we can get the derivatives:

$$\frac{\partial \mathcal{L}}{\partial w_j} = \frac{1}{N} \sum_{i=1}^{N} x_j^{(i)} \left(\sum_{j'} w_{j'} x_{j'}^{(i)} + b - y^{(i)} \right) = \frac{1}{N} \sum_{i=1}^{N} x_j^{(i)} \left(\hat{y}^{(i)} - y^{(i)} \right)$$

$$\frac{\partial \mathcal{L}}{\partial b} = \frac{1}{N} \sum_{i=1}^{N} \left(\sum_{j'} w_{j'} x_{j'}^{(i)} + b - t^{(i)} \right) = \frac{1}{N} \sum_{i=1}^{N} \hat{y}^{(i)} - y^{(i)}. \tag{3.5}$$

The derivatives are simplified a bit. Basically, there are two ways to use these partial derivatives to solve this regression problem.

1. Direct solution

One way to compute the minimum of a function is to set the partial derivatives to zero. Recall from single-variable calculus that (assuming a function is differentiable) the minimum $x^\star$ of a function f has the property that the derivative df/dx is zero at $x = x^\star$. Note that the converse is not true: if $df/dx = 0$, then $x^\star$ might be a maximum or an inflection point, rather than a minimum. But the minimum can only occur at points that have a zero derivative. An analogous result holds in the multivariate case: If f is differentiable, then all of the partial derivatives $\partial f/\partial x_j$ are zero at the minimum. The intuition is simple: if $\partial f/\partial x_j$ is positive, then one can decrease f slightly by decreasing x_j slightly. Conversely, if $\partial f/\partial x_j$ is negative, then one can decrease f slightly by increasing x_j slightly. In either case, this implies we're not at the minimum. Therefore, if the minimum exists (i.e., f doesn't keep growing as x goes to infinity), it occurs at a critical point, i.e., a point where the partial derivatives are zero. This gives us a strategy for finding minima: Set the partial derivatives to zero, and solve for the parameters. This method is known as direct solution.

Let's apply this to linear regression. For simplicity, let's assume the model doesn't have a bias term. (We actually don't lose anything by getting rid of the bias. Just add a "dummy" input x_0 that always takes the value 1; then, the weight w_0 acts as a bias.) We simplify Eq. (3.5) to remove the bias, and set the partial derivatives to zero:

$$\frac{\partial \mathcal{L}}{\partial w_j} = \frac{1}{N} \sum_{i=1}^{N} x_j^{(i)} \left(\sum_{j'=1}^{M} w_{j'} x_{j'}^{(i)} - y^{(i)} \right) = 0. \qquad (3.6)$$

In this linear equation, we have N data samples and M weights. As long as $N \geq M$, this system of linear equations can be solved. Since we're trying to solve for the weights, let's pull these out:

$$\frac{\partial \mathcal{L}}{\partial w_j} = \frac{1}{N} \sum_{j'=1}^{M} \left(\sum_{i=1}^{N} x_j^{(i)} x_{j'}^{(i)} \right) w_{j'} - \frac{1}{N} \sum_{i=1}^{N} x_j^{(i)} y^{(i)} = 0. \qquad (3.7)$$

You can observe that this is a system of M linear equations in M variables that can be further simplified as

$$\sum_{j'=1}^{M} A_{jj'} w_{j'} - c_j = 0 \quad \forall j \in \{1, \ldots, M\}, \qquad (3.8)$$

where $A_{jj'} = \frac{1}{N} \sum_{i=1}^{N} x_j^{(i)} x_{j'}^{(i)}$ and $c_j = \frac{1}{N} \sum_{i=1}^{N} x_j^{(i)} t^{(i)}$. Until now, we are able to solve the linear regression equations by computing all the values $A_{jj'}$ and c_j.

Note that the solution we just derived is very particular to linear regression. In general, the system of equations will be nonlinear, and, except in rare cases, systems of nonlinear equations don't have closed-form solutions. Linear regression is very unusual in that it has a closed-form solution.

2. Gradient descent

Now, let's minimize the cost function in another way: gradient descent. This is an iterative algorithm, which means that we apply a certain update rule over and over again, and, if we're lucky, our iterates will gradually improve according to our objective function. To do gradient descent, we initialize the weights to some value (e.g., all zeros), and repeatedly adjust them in the direction that most decreases the cost function. If we visualize the cost function as a surface, so that lower is better, this is the direction of steepest descent. We repeat this procedure until the iterates converge, or stop changing much. If we're lucky, the final iterate will be close to the optimum.

In order to make this mathematically precise, we must introduce the **gradient**, the direction of steepest ascent (i.e., fastest increase) of a function. The entries of the gradient vector are simply the partial derivatives with respect to each of the variables as shown in the following example:

$$\frac{\partial \mathcal{L}}{\partial \mathbf{w}} = \begin{pmatrix} \frac{\partial \mathcal{L}}{\partial w_1} \\ \vdots \\ \frac{\partial \mathcal{L}}{\partial w_M} \end{pmatrix}. \tag{3.9}$$

This formula gives the direction of steepest ascent. It also suggests that, to decrease a function as quickly as possible, we should update the parameters in the direction opposite the gradient.

Thus, we can formalize the function-decreasing process using the following update rule, which is known as gradient descent:

$$\mathbf{w} \leftarrow \mathbf{w} - \alpha \frac{\partial \mathcal{L}}{\partial \mathbf{w}}, \tag{3.10}$$

or in terms of coordinates,

$$w_j \leftarrow w_j - \alpha \frac{\partial \mathcal{E}}{\partial w_j}. \tag{3.11}$$

The symbol $\leftarrow$ means that the left-hand side is updated to take the value on the right-hand side; the constant α is known as a learning rate. The larger it is, the larger a step we take. But, in general, it's good to choose a small value such as 0.01 or 0.001. If we plug in the formula into Eq. (3.11) for the partial derivatives of the regression model, we get the update rule, which is also the core of neural network updating rules:

$$w_j \leftarrow w_j - \alpha \frac{1}{N} \sum_{i=1}^{N} x_j \left(\hat{y}^{(i)} - y^{(i)} \right). \tag{3.12}$$

We just need to repeat this update many times until the gradient goes close to zero. You might ask: By setting the partial derivatives to zero, do we actually compute the exact solution? With gradient descent, we never actually reach the optimum, but merely approach it gradually. Then, why would we ever prefer gradient descent? There are two reasons:

- We can only solve the system of equations explicitly for a handful of models. By contrast, we can apply gradient descent to any model for which we can compute the gradient. This is usually pretty easy to do efficiently. Importantly, it can usually be done automatically, so software packages like PyTorch and TensorFlow can save us from ever having to compute partial derivatives by hand.
- Solving a large system of linear equations can be expensive, possibly many orders of magnitude more expensive than a single gradient-descent update.

Therefore, gradient descent can sometimes find a reasonable solution much faster than solving the linear system. Therefore, gradient descent is often more practical than computing exact solutions, even for models where we are able to derive the latter. For these reasons, gradient descent will be our workhorse throughout the course. We will use it to train almost all our models, with the exception of a handful for which we can derive exact solutions.

Vectorization

We can also introduce linear algebra to rewrite the linear regression model, as well as both solution methods, in terms of operations on matrices and vectors. This process is known as vectorization. Writing those equations using matrices and vectors has at least two advantages:

- Vector-formed formulas can be much simpler, more compact, and more readable in this form.
- Vectorized code can be much faster than explicit for-loops, for several reasons:
  - High-level programming languages like Python can introduce a lot of interpreter overhead, and, if we explicitly write a for-loop, it might be $10 - 100$ times slower than the C programming language equivalent. If we instead write the algorithm in terms of a much smaller number of linear algebra operations, then it can perform the same computations much faster with minimal interpreter overhead.
  - Since linear algebra is used all over the place, linear algebra libraries have been extremely well optimized for various computer architectures. Hence, they use much more efficient memory access patterns than a naive for-loop.
  - Matrix multiplication is inherently highly parallelizable and involves little control flow. Hence, it's ideal for graphics processing unit (GPU) architectures. If you run vectorized code on a GPU using a framework like TensorFlow or PyTorch, it may run 50 times faster than the CPU version. As it turns out, most of the computation in deep learning involves matrix multiplication, which is why it's been such an incredibly good match for GPUs.

In general, matrices will be denoted with capital boldface, vectors with lowercase boldface, and scalars with plain type. If we have N training examples, each M-dimensional, we will represent the inputs as an $N \times M$ matrix $\mathbf{X}$, i.e., $\mathbf{X} \in \mathbb{R}^{N \times M}$. Each row of $\mathbf{X}$ corresponds to a training example, and each column corresponds to a single input dimension. The weights are represented as a M-dimensional vector $\mathbf{w}$, and the targets are represented as a N-dimensional vector $\mathbf{t}$. The predictions are computed using a matrix-vector product

$$\hat{\mathbf{y}} = \mathbf{X}\mathbf{w} + b\mathbf{1}, \tag{3.13}$$

where **1** denotes a vector of all ones. We can express the cost function in vectorized form:

$$\mathcal{L} = \frac{1}{2N}\|\hat{\mathbf{y}} - \mathbf{y}\|^2$$
$$= \frac{1}{2N}\|\mathbf{X}\mathbf{w} + b\mathbf{1} - \mathbf{y}\|^2. \tag{3.14}$$

Note that this is considerably simpler than Eq. (3.3). Even more importantly, it saves us from having to explicitly sum over the indices i and j. As our models get more complicated, we would run out of convenient letters to use as indices if we didn't vectorize those variables.

As shown in Eq. (3.8), we derived a system of linear equations, with coefficients $A_{jj'} = \frac{1}{N}\sum_{i=1}^{N} x_j^{(i)} x_{j'}^{(i)}$ and $c_j = \frac{1}{N}\sum_{i=1}^{N} x_j^{(i)} y^{(i)}$. Using linear algebra, we can write these as the matrix $\mathbf{A} = \frac{1}{N}\mathbf{X}^\top\mathbf{X}$ and $\mathbf{c} = \frac{1}{N}\mathbf{X}^\top\mathbf{y}$. The solution to the linear system $\mathbf{A}\mathbf{w} = \mathbf{c}$ is given by $\mathbf{w} = \mathbf{A}^{-1}\mathbf{c}$ (assuming $\mathbf{A}$ is invertible), and, in this way, the optimal weights can be directly written in a vectorized form:

$$\mathbf{w} = \left(\mathbf{X}^\top\mathbf{X}\right)^{-1}\mathbf{X}^\top\mathbf{y}. \tag{3.15}$$

An exact solution that we can express with a formula is known as a closed-form solution.

Similarly, we can vectorize the gradient descent update process from Eq. (3.12)

$$\mathbf{w} \leftarrow \mathbf{w} - \frac{\alpha}{N}\mathbf{X}^\top(\hat{\mathbf{y}} - \mathbf{y}), \tag{3.16}$$

where $\hat{\mathbf{y}}$ is computed as in Eq. (3.13).

3.2.2 Logistic regression

Logistic regression is one popular binary classification model. Logistic regression is actually the simplest neural network model, which is also known as the perceptron. Next, let us explain logistic regression with a traffic example.

Suppose we want to determine whether a road segment is congested or not based on the traffic speed, volume, and other physical attributes of the road segment. In this example, the label $y = 1$ if the segment is congested and $y = 0$ otherwise. Each road segment has a M dimensional feature vector x.

Mathematically logistic regression models the probability of congestion $y = 1$ given input features x, denoted by $P(y = 1 \mid x)$. Then, the classification is performed by comparing $P(y = 1 \mid x)$ with a threshold (e.g., 0.5). If $P(y = 1 \mid x)$ is greater than the threshold, we predict the road segment will is congested; otherwise, the road segment is not.

One building block of logistic regression is the log-odds or logit function. The odds is the quantity that measures the relative probability of label presence and label absence as

$$\frac{P(y = 1 \mid x)}{1 - P(y = 1 \mid x).}$$

The lower the odds, the lower probability of the given label. Sometimes, we prefer to use log-odds (natural logarithm transformation of odds), also known as the logit function:

$$logit(x) = \log \left(\frac{P(y = 1 \mid x)}{1 - P(y = 1 \mid x)} \right). \tag{3.17}$$

Now, instead of modeling the probability of road segment congestion label given input feature $P(y = 1 \mid x)$ directly, it is easier to model its logit function as a linear regression over x:

$$\log \left(\frac{P(y = 1 \mid x)}{1 - P(y = 1 \mid x)} \right) = w^T x + b, \tag{3.18}$$

where w is the weight vector and b is the offset variable. Eq. (3.18) is why logistic regression is named logistic regression.

After taking exponential to both sides and some simple transformation, we will have the following formula:

$$P(y = 1 \mid x) = \frac{e^{w^T x + b}}{1 + e^{w^T x + b}}. \tag{3.19}$$

With the formulation in Eq. (3.19), the logistic regression will always output values between 0 and 1, which is desirable for probability estimation.

Let us denote $P(y = 1 \mid x)$ as $P(x)$ for simplicity. Now, learning the logistic regression model means estimating the parameters w and b on the training data. We often use maximum likelihood estimation (MLE) to find the parameters. The idea is to estimate w and b so that the prediction $\hat{P}(x_i)$ to data point i in the training data is as close as possible to the actual observed values (in this case, either 0 or 1). Let x_+ be the set of indices for data points that belong to the positive class (i.e., with road congestion) and x_- one be the set of data points that belong to the negative class (i.e., without road congestion). Then, the likelihood function used in the MLE is given by:

$$\mathcal{L}(\mathbf{w}, b) = \prod_{a_+ \in \mathbf{x}_+} P\left(\mathbf{x}_{a_+}\right) \prod_{a_- \in \mathbf{x}_-} \left(1 - P\left(\mathbf{x}_{a_-}\right)\right). \tag{3.20}$$

If we apply the logarithm to the MLE, we will get the following formula for log-likelihood:

$$\log(\mathcal{L}(\mathbf{w}, b)) = \sum_{i=1}^{N} \left[y_i \log P\left(x_i\right) + (1 - y_i) \log \left(1 - P\left(x_i\right)\right) \right]. \tag{3.21}$$

Note that, since either y_i or $1 - y_i$ is zero, only one of two probability terms (either $\log P(x_i)$ or $\log(1 - P(x_i))$) will be added.

Multiplying a negative sign to have a minimization problem, what we have now is the negative log-likelihood, also known as (binary) **cross-entropy loss**:

$$J(\mathbf{w}, b) = -\sum_{i=1}^{N} \left[y_i \log P(\mathbf{x}_i) + (1 - y_i) \log(1 - P(\mathbf{x}_i)) \right]. \tag{3.22}$$

Maximizing the log-likelihood is identical to minimizing the cross-entropy loss. In this setting, we can use the gradient descent to find the optimal weights w and b.

Softmax regression

Logistic regression is designed for binary classification problems. We sometimes want to classify data points into more than two classes, e.g., given traffic-sign image data, such as stop, right turn, and left turn. In that case, we will use multinomial logistic regression, also called softmax regression to model this problem.

Assuming we have K classes, the goal is to estimate the probability of the class label taking on each of the K possible categories $P(y = k \mid x)$ for $k = 1, \cdots, K$. Thus, we will output a K-dimensional vector representing the estimated probabilities for all K classes. The probability that data point i is in class a can be modeled as:

$$P(y_i = a \mid \mathbf{x}_i) = \frac{e^{\mathbf{w}_a^T \mathbf{x}_i + b_a}}{\sum_{k=1}^{K} e^{\mathbf{w}_k^T \mathbf{x}_i + b_k}}, \tag{3.23}$$

where $\mathbf{w}_a$ is the weight for a-th class, $\mathbf{x}_i$ is the feature vector for data point i, and b_a is the offset for class a, respectively. To learn parameters for softmax regression, we can optimize the following average cross-entropy loss over all N training data points:

$$J(w) = -\frac{1}{N} \sum_{i=1}^{N} \sum_{k=1}^{K} I(y_i = k) \log(P(y_i = k \mid x_i)), \tag{3.24}$$

where K is the number of label classes (e.g., three classes in traffic-sign image classification), $I(y_i = k)$ is binary indicator (0 or 1) if k is the class for data point i, and $P(y_i = k \mid \mathbf{x}_i)$ is the predicted probability that data point i is of class k. In this way, softmax regression is just a multi-class extension of the logistic regression.

3.3 Unsupervised learning

In many travel behavior research problems, we can hardly predefine the categories of those behaviors. In such cases, we resort to unsupervised learning models. Unsupervised learning models are not used for classifying (or predicting) towards a known label y. Rather, they discover patterns or clusters about the input data x. Next, we briefly introduce some popular unsupervised learning methods.

3.3.1 Principal component analysis

Suppose we want to study N data points of M features represented by a matrix $X \in \mathbb{R}^{N \times M}$. The number of features M can be large in many travel behavior or trajectory datasets. For example, the dimension of a passenger's travel behavior data over a week can reach 24*7 if we simply sample the location from the passenger's trajectory every hour. Principal component analysis (PCA) can be applied to reduce the data dimensionality from M to a much lower dimension R. More specifically, PCA is the linear transformation:

$$Y = XW, \tag{3.25}$$

where X is the original data matrix, $Y \in \mathbb{R}^{N \times R}$ is the low-dimensional representation after PCA, and $W \in \mathbb{R}^{M \times R}$ is the orthogonal projection matrix. The objective of PCA is to minimize the reconstruction error:

$$\min_{W} \left\| X - XWW^{\top} \right\|^{2}, \tag{3.26}$$

where $XWW^{\top} = YW^{\top}$ is the reconstruction matrix. The solution of PCA relates to another matrix factorization named singular value decomposition (SVD):

$$X \approx U\Sigma W^{\top}, \tag{3.27}$$

where $U \in \mathbb{R}^{M \times R}$ and $W \in \mathbb{R}^{N \times R}$ are orthogonal matrices [2] that contain left and right singular vectors and $\Sigma \in \mathbb{R}^{R \times R}$. Connecting to PCA, if X is the high-dimensional data matrix, the low-dimensional representation is $Y = U\Sigma = XW$.

In most cases, PCA are combined with other methods and used as a feature-extraction method for generating features from high-dimensional data such as image data. For example, in the transportation domain, image data such as vehicle or passengers includes many different appearances in pixel level, whose high dimensionality causes many challenges for traffic object detection and classification tasks. PCA can be used to extract the high-level feature to assist detection accuracy. To summarize, as an unsupervised learning method, PCA provides low-dimensional linear representation to approximate the original high-dimensional features. In fact, PCA can also be achieved by a neural network via autoencoders with linear activation if we have enough data to train the neural network.

3.3.2 Clustering

Besides dimensionality reduction, clustering is another major topic in unsupervised learning. Clustering methods aim to find homogeneous groups (or clusters) from a dataset, such that similar points are within the same cluster but dissimilar points are in different clusters. For example, researchers have applied clustering algorithms on transit passenger's travel behavior data to find abnormal travel behaviors and to categorize the transit passengers into various groups, such as early birds, night owls, tireless itinerants, and recurring itinerants (Cui and Long, 2019).

One popular clustering method is K-means, which tries to group data into K clusters where users specify the number K. The clustering assignments are achieved by minimizing the sum of distances (e.g., Euclidean distances) between data points and the corresponding cluster centroid (or the mean vectors). The K-means method is described in the following section.

3.4 Key concepts in machine learning

3.4.1 Loss

Loss functions are used to determine the error (aka "the loss") between the output of our algorithms and the given target value. This expresses how far off the mark our computed output is.

There are multiple ways to determine loss. Two of the most popular loss functions in machine learning are the 0–1 loss function and the quadratic loss function. The 0–1 loss function is an indicator function that returns 1 when the target and output are not equal and zero otherwise.

The quadratic loss is a commonly used symmetric loss function. The quadratic losses' symmetry comes from its output being identical in relation to targets that differ by some value x in any direction (i.e., if the output overshoots by 1, that is the same as undershooting by 1).

In summary, loss functions are used in optimization problems with the goal of minimizing the loss. In simple linear regression tasks, we can find the line of best fit by minimizing the overall loss of all the points with the prediction from the line. Loss functions are used to train neural networks by influencing how their weights are updated. The larger the loss is, the larger the update. By minimizing the loss and updating the model, the model's accuracy is maximized. However, the tradeoff between the size of the update and the minimal loss must be evaluated in these machine learning applications.

3.4.2 Regularization

In regression analysis, the features are estimated using coefficients/weights while modeling. If the weights of the features can be restricted, shrunk, or regularized towards zero, the impact of insignificant features might be reduced in the model. In such a way, we can achieve a stable fit of the model and prevent the model having high variance.

Regularization is the most-used technique to penalize complex models in machine learning. It assumes that least weights may produce simpler models and hence assist in avoiding overfitting. Regularization is deployed by penalizing the model weights of high values. It can enhance the performance of models for new inputs. Since it makes the magnitude to weighted values low in a model, regularization technique is also referred to as weight decay.

In general, regularization is adopted universally because simple data models generalize better and are less prone to overfitting. Examples of regularization include:

- Restricting the segments for avoiding redundant groups in K-means;
- Confining the complexity (weights) of neural networks;
- Reducing the depth of tree and branches (new features) in random forests.

Although there are various regularization techniques, the most well-known techniques are L1 and L2 regularization. Both L1 and L2 can add a penalty to the cost depending upon the model complexity, so that, at the place of computing the cost by using a loss function, there will be an auxiliary component, known as the regularization terms, added in order to penalizing complex models. By adding regularization terms, the value of weights matrices is reduced by assuming that a neural network having less weights makes simpler models. Hence, it reduces the overfitting to a certain level.

- L1 regularization: It adds an L1 penalty that is equal to the absolute value of the magnitude of coefficient, or simply restricting the size of coefficients. L1 regularization is the preferred choice when having a high number of features because it provides sparse solutions. Moreover, we obtain the computational advantage because features with zero coefficients can be avoided. For example, Lasso regression implements this method.
- L2 Regularization: This adds an L2 penalty that is equal to the square of the magnitude of coefficients. L2 regularization can deal with the multicollinearity (independent variables are highly correlated) problems through constricting the coefficient and by keeping all the variables. For example, Ridge regression and SVM employ this method.
- Elastic Net: When L1 and L2 regularization are combined, it becomes the elastic net method and adds a hyper-parameter.

L1 vs. L2

It is often observed that people get confused in selecting the suitable regularization approach to avoid overfitting while training a machine learning model. Among many regularization techniques, such as L2 and L1 regularization, dropout, data augmentation, and early stopping, we will learn here intuitive differences between L1 and L2 regularization.

1. Where L1 regularization attempts to estimate the median of data, L2 regularization makes estimates for the mean of the data in order to evade overfitting.
2. Through including the absolute value of weight parameters, L1 regularization can add the penalty term in the cost function. On the other hand, L2 regularization appends the squared value of weights in the cost function.
3. Feature selection is the characteristic of holding highly significant coefficients, either very close to zero or not very close to zero, where in general coefficients approaching zero would be eliminated later. This will lead to the sparsity where certain features are expels from the model. In this context, L1 regularization can be helpful in features selection by eradicating the unimportant features, whereas L2 regularization is not recommended for feature selection.
4. L2 has a solution in closed form since it's a square of a weight; on the other side, L1 doesn't have a closed-form solution since it includes an absolute value and is a non-differentiable function. Due to this reason, L1 regularization is relatively more expensive in computation, can't be solved in the context of matrix measurement, and heavily relies on approximations.

3.4.3 Gradient descent vs. gradient ascent

Gradient descent and gradient ascent are equivalent from the perspective of mathematical representation. Gradient ascent is just the process of maximizing an objective function, instead of minimizing a loss function. Everything else is entirely the same. Ascent for some loss function is like gradient descent on the negative of that loss function.

1. Gradient Descent: gradient descent is minimizing the cost function used in linear regression, and it provides a downward or decreasing slope of cost function. If you want to minimize a function, use the gradient descent. For example, we want to minimize the loss function in deep learning, so we use the gradient descent.
2. Gradient Ascent: gradient ascent is maximizing a function so as to achieve better optimization. For example, in reinforcement learning the goal of policy gradient methods is to maximize the reward/expected return function, so we use the gradient ascent.

3.4.4 K-fold cross-validation

K-fold cross-validation is a technique to compare and select a model when training a model. When the training data is scarce, we might not be able to afford to withhold enough data to constitute a proper validation set. One popular solution to this problem is to employ K-fold cross validation, in which the original training data is split into K nonoverlapping subsets. In general, the K-fold cross-validation procedure can be described as follows:

1. Shuffle the dataset randomly.

2. Split the dataset into K groups.
3. For each unique group:
    a. Take the group as a hold out or test data set;
    b. Take the remaining groups as a training data set;
    c. Fit a model on the training set and evaluate it on the test set;
    d. Retain the evaluation score and discard the model.
4. Estimate the training and validation errors by averaging the results from the K experiments.

3.5 Exercises

3.5.1 Questions

1. What is the key difference between supervised learning and unsupervised learning?
2. What is the difference between the tasks fulfilled by linear regression and logistic regression?
3. How does linear regression find the optimal point?
4. What are the loss functions used in linear regression?
5. Why can't we use the mean square error cost function used in linear regression for logistic regression?
6. What are the differences between L1 and L2 regularization terms?
7. The logit function is defined as the log of the odds function. What do you think the input range of this logit function be in the domain of [0, 1]?
    a. (−infinity, +infinity);
    b. (0, +infinity);
    c. (−infinity, 0);
    d. (0, 1).
8. What is the underlying method that is used to fit the training data in the algorithm of logistic regression?
    a. Jaccard distance;
    b. Maximum likelihood;
    c. Least square error;
    d. None of the options just mentioned.
9. Which strategy is not for avoiding overfitting?
    a. Adding regularization term in the loss function;
    b. Making a more complex model;
    c. Using cross-validation like k-folds;
    d. Using a validation set while training.

Chapter 4

Fully connected neural networks

Following a brief introduction of machine learning basics, this chapter will introduce a type of neural network with the most basic structure, i.e., the fully connected neural network or feed-forward neural network (FNN). When people say neural network, they are normally referring to FNN. FNN with fully connections between neurons in different layers of the neural network can solve various tasks, like regression and classification tasks. In addition, fully connected neural network layers are the basic elements of deep neural networks. Thus, in this chapter we will introduce the basic structure of FNNs starting from linear regression to linear neural networks. Key components of deep neural networks, including layers, activation functions, and loss functions, are introduced. The back-propagation method as the core of neural network training is also elaborated to shed light on the power of deep neural networks. Subsequently, this chapter demonstrates transportation applications that can be easily solved by FNN.

4.1 Linear regression

In order to make things easier to grasp, we begin with the simplest concepts. We will start from classic algorithms and provide the basis for more complex techniques in the rest of the book.

In a linear model, the target can be expressed as a weighted sum of the features. For example, traffic volume can be expressed as a weighted sum of a number of lanes and time:

$$\text{volume} = w_{\#\text{ of lanes}} \cdot \#\text{ of lanes} + w_{\text{time}} \cdot \text{time} + b. \qquad (4.1)$$

$w_{\#\text{ of lanes}}$ and w_{time} are called weights, and b is called a bias (also called an offset or intercept) in a linear model. The weights determine the influence of each feature on the prediction/output (volume), and the bias indicates the predicted volume value when all of the features take value 0. We need the bias to increase the expressivity of the model. Given a dataset, the goal of the linear model is to choose the weights and the bias such that, on average, the predictions best fit the true volume observed in the data.

In machine learning, we usually process high-dimensional data whose linear algebra notation will be most convenient for us to write the equation in a simpler

Machine Learning for Transportation Research and Applications
https://doi.org/10.1016/B978-0-32-396126-4.00009-6

way. When our inputs consist of d features, the prediction $\hat{y}$ (the "hat" symbol $\hat{y}$ generally denotes predictions/estimates) can be expressed as

$$\hat{y} = w_1 x_1 + \ldots + w_d x_d + b. \tag{4.2}$$

Representing all features as a vector $\mathbf{x} \in \mathbb{R}^d$ and all weights as a vector $\mathbf{w} \in \mathbb{R}^d$, the linear model can be described compactly using a dot product:

$$\hat{y} = \mathbf{w}^\top \mathbf{x} + b. \tag{4.3}$$

The vector $\mathbf{x} \in \mathbb{R}^d$ corresponds to features of a single data example (we also call it a sample). The features of the entire dataset with n samples can be represented by a matrix $\mathbf{X} \in \mathbb{R}^{n \times d}$. Here, $\mathbf{X}$ contains one row for every sample and one column for every feature. In this way, the predictions $\hat{\mathbf{y}} \in \mathbb{R}^n$ can be expressed via the matrix-vector product:

$$\hat{\mathbf{y}} = \mathbf{X}\mathbf{w} + b. \tag{4.4}$$

In the linear regression setting, Eq. (4.4) has an analytic solution with a simple formula:

$$\mathbf{w}^* = \left(\mathbf{X}^\top \mathbf{X} \right)^{-1} \mathbf{X}^\top \mathbf{y}. \tag{4.5}$$

Although simple problems like linear regression may have analytic solutions, it cannot be applied in deep learning because the requirement of an analytic solution is too restrictive to be satisfied.

In the machine learning/deep learning setting, the goal of Eq. (4.4) is to fit a good model with the parameters $\mathbf{w}$ and b based on a training dataset $\mathbf{X}$ and the corresponding training labels $\mathbf{y}$. When testing the model, the predicted value $\hat{\mathbf{y}}$ will generate the lowest error. One thing to note is that, when we assume the best model is a linear model, we cannot expect the prediction error $||y_i - \hat{y}_i||$ for the data samples $\mathbf{X}_i$ to be exactly zero. Thus, we will incorporate a noise term to account for such errors.

In the deep learning setting, we also have the training dataset and testing dataset. It is important to figure out how to measure the quality of the model and how to improve the model during the training procedure, which requires the loss function and back-propagation, respectively, which are introduced in the following sections.

4.2 Deep neural network fundamentals

4.2.1 Perceptron

A neural network, as a branch of artificial intelligence methods, has been developed and applied to many problems for a considerable time. Perceptron Rosenblatt (1957) might be the oldest neural network still in use today. It is

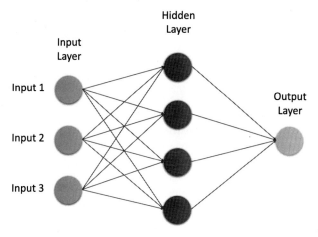

FIGURE 4.1   Multilayer perceptron.

also the simplest fully connected neural network. A very common perceptron method is the multilayer perceptron (MLP), normally with three layers, an input layer, a hidden layer, and an output layer, as shown in Fig. 4.1, where the nodes/neurons between layers are all connected without loops. In the hidden layer, the node accepts multiple inputs, each input is multiplied by a weight, and the products are added up, which is similar to the procedure in linear regression. The difference is that, in the neural networks, there are activation functions between layers. Neural networks are simulations of biological neurons, where the weights simulate the role of synapses in biological neurons to enhance or inhibit a signal. The result, after adding a bias value, will act as input in an activation function before it is passed to the next layer. An activation function simulates the neuron firing or not. For example, in a binary activation function, if the sum of weighted inputs and bias is greater than zero, the neuron output is 1 (it fires), otherwise 0 (it does not fire).

A single layer with limited neurons cannot solve the problems when the data are nonlinearly separable. Adding one (or more) hidden layers makes it possible to solve those problems. Thus, the multilayer structure empowers the neural networks to process data with nonlinear patterns. Per the universal approximation theorem, an FNN with one hidden layer can represent any function (Cybenko, 1989), although, in the early stage training, such a multilayer model is very difficult without high performance computation devices, like a graphical processing unit (GPU).

Given activation functions and multiple layers, the multilayer FNN still cannot work without a learning algorithm, intended to improve the model's prediction performance while training. In the 1980s, the back-propagation (BP) algorithm enabling learning in a multilayer FNN was proposed (Rumelhart et al., 1986), and it soon resulted in a renewed interest in the field. In the last decade, with the help of the BP algorithm and advanced GPU technologies, we witnessed the rise of deep learning. Training complex neural networks is no longer a big problem to develop a variety of deep neural networks to solve very complicated problems.

4.2.2 Hidden layers

A multilayer neural network consists of an input layer, an output layer, and one or more hidden layers (between the input and output layers). One of the obvious advantages of a multilayer neural network is the capability to model the complex mapping relationship between the input and output data. Incorporating one or more hidden layers can help the neural network to handle a more general class of functions.

The easiest neural network architecture is stacking many fully connected layers on top of each other, where each layer feeds into the layer above it, until the neural network generates outputs. In this way, we can consider the first $L - 1$ layers as the extracted representation of the data and the final layer as a linear predictor. Taking the MLP as an example, as shown in Fig. 4.1, this MLP has four inputs, three outputs, and its hidden layer contains five hidden units. Since the input layer does not involve any calculations, the number of fully connected layers in this MLP is two because producing outputs with this network requires implementing the computations for both the hidden and output layers. Every input element is connected to every neuron in the hidden layer, and each of these hidden neurons in turn influences every neuron in the output layer.

This basic fully connected layer can be formulated via mathematical equations. The input data matrix $\mathbf{X} \in \mathbb{R}^{n \times d}$ has two dimensions, where n denote the batch size that $\mathbf{X}$ has n sample data/examples where each example has d inputs (features). For a one-hidden-layer MLP whose hidden layer has h hidden neurons/units, the outputs of the hidden layer, i.e., the hidden representations, can be denoted as $\mathbf{H} \in \mathbb{R}^{n \times h}$. $\mathbf{H}$ is also known as a hidden-layer variable or a hidden variable. Since the hidden and output layers are both fully connected, we have the hidden-layer weights $\mathbf{W}^{(1)} \in \mathbb{R}^{d \times h}$ and biases $\mathbf{b}^{(1)} \in \mathbb{R}^{1 \times h}$ and output-layer weights $\mathbf{W}^{(2)} \in \mathbb{R}^{h \times k}$ and biases $\mathbf{b}^{(2)} \in \mathbb{R}^{1 \times k}$. Formally, we calculate the outputs $\mathbf{Y} \in \mathbb{R}^{n \times k}$ of the one-hidden-layer MLP as follows:

$$\begin{aligned} \mathbf{H} &= \mathbf{X}\mathbf{W}^{(1)} + \mathbf{b}^{(1)} \\ \mathbf{Y} &= \mathbf{H}\mathbf{W}^{(2)} + \mathbf{b}^{(2)}. \end{aligned} \tag{4.6}$$

Note that, after adding the hidden layer, the neural network has more weight parameters to track and update, but whether the neural network's capability of modeling the data has increased is not apparent. You might be surprised to find out that the answer is NO! The reason is that the hidden units defined in Eq. (4.6) are given by an affine function of the inputs, and the outputs (without softmax) are just an affine function of the hidden units. Please note that an affine function of an affine function is itself an affine function. Moreover, a linear model has already been able to represent any affine function. We can view the equivalence formally by proving that, for any values of the weights, we can just collapse out the hidden layer, yielding an equivalent single-layer linear model with parameters:

$$Y = \left(\mathbf{X}\mathbf{W}^{(1)} + \mathbf{b}^{(1)}\right)\mathbf{W}^{(2)} + \mathbf{b}^{(2)}$$

$$Y = \mathbf{X}\underbrace{\mathbf{W}^{(1)}\mathbf{W}^{(2)}}_{W} + \underbrace{\mathbf{b}^{(1)}\mathbf{W}^{(2)} + \mathbf{b}^{(2)}}_{b} \tag{4.7}$$

$$Y = \mathbf{X}\mathbf{W} + \mathbf{b}.$$

4.2.3 Activation functions

As Eq. (4.7) states, the combination of affine functions are still linear functions. Thus, we need to add the nonlinear property into the neural network to build the complex mapping relationship between its input and output. Since neural networks has a multilayer structure, the nonlinear function can be added between different layers, which originated from the role of synapses in biological neurons to enhance or inhibit a signal. The nonlinear activation function σ is normally applied to the affine function such that a layer with activation function can be described as

$$Y = \sigma(\mathbf{X}\mathbf{W} + \mathbf{b}). \tag{4.8}$$

The activation functions σ are differentiable operators to transform input signals to outputs, where the input signals will be adjusted or rescaled to simulate the enhancing or inhibiting of the signals. It usually maps the resulting values in between 0 to 1 or -1 to 1, like determining whether to activate the neurons of a neural network layer or not. There several classical and widely used activation functions, like sigmoid, tanh, relu. In the following subsections, we will briefly introduce several common and widely used activation functions.

Sigmoid function

The sigmoid/logistic function curve looks like an S-shape, in which it transforms the inputs from the real value domain $\mathbb{R}$ to $(0, 1)$:

$$\text{sigmoid}(x) = \frac{1}{1 + \exp(-x)}. \tag{4.9}$$

Sigmoid activation function has been widely used for several reasons:

1. Since the probability of anything existing can only be in the range of 0–1, sigmoid looks like it outputs the probability of whether a neuron of the neural network is activated.
2. The sigmoid function is differentiable and provides a smooth gradient, i.e., preventing jumps in output values.

However, sigmoid also has its disadvantages in that it can cause a neural network to get stuck during the training procedure. The reason will be discussed in the following sections.

Tanh function

The hyperbolic tangent (tanh) activation function looks like the sigmoid function with s-shape. The range of the tanh function is from -1 to 1:

$$\tanh(x) = \frac{2}{1 + e^{-2x}} - 1. \tag{4.10}$$

The tanh activation function is differentiable and monotonic, while its derivative is not monotonic. It is mainly used for classification between two classes. Both tanh and sigmoid functions are differentiable, but they have the potential to result in vanishing gradient during the training process, which will be addressed later.

ReLU function

ReLU is probably the most used activation function in the world right now. ReLU and its variants are used in all kinds of neural networks. It does not map the input to a limited range, and its function is pretty simply that

$$\text{ReLU}(x) = \max(0, x). \tag{4.11}$$

As the equations shows, the activated value is zero when x is less than zero and equals to itself when x is above or equal to zero. The ReLU function and its derivative both are monotonic. Further, its derivatives are particularly well behaved: Either they vanish or they just let the argument through. This simple structure brings several advantages such that ReLU needs less time and space complexity for calculation. It also mitigates the vanishing gradient problem. However, it does not avoid the gradient exploding problem. It also introduces the dying ReLU problem when the learning rate is too high or there is a large negative bias.

Therefore, there are many variants of the ReLU function, such as the Leaky ReLU, which is the most common and effective method to solve a dying ReLU problem. Leaky ReLU adds a slight slope in the negative range to prevent the dying ReLU problem.

4.2.4 Loss functions

A loss function is a simple concept that evaluates how well the algorithm models the dataset. Specifically, it is calculated to determine the error (loss) between the output of our algorithms and the given target value. The loss function calculates the loss that is usually a nonnegative number and where smaller values are better and perfect predictions incur a loss of 0. In machine learning or deep learning algorithms, there are multiple ways to formulate the loss functions depending on the specific learning tasks.

The most popular loss function in regression problems is the mean squared error. When our prediction for an sample data entry i is $\hat{y}_i$ and the corresponding true label is y_i, the squared error is given by:

$$L(\hat{y}_i, y_i) = \frac{1}{n} \sum_{i=1}^{n} (\hat{y}_i - y_i)^2. \tag{4.12}$$

The loss function is used to optimize the weights of a model to achieve a smaller loss when we train the model.

Taking the one-layer neural network described in Eq. (4.8) as an example, the loss function is calculated with respect to the weights $\mathbf{W}$ and $\mathbf{b}$. Thus, a neural network targeted to minimize the loss can be optimized based on the following formulation to acquire a set of optimal or near optimal weights:

$$\mathbf{W}^*, \mathbf{b}^* = \underset{\mathbf{W}, \mathbf{b}}{\arg\min} \, L(\hat{y}, y). \tag{4.13}$$

There are various forms of loss functions depending on specific tasks. Classification problems usually incorporate the cross entropy loss. When we attempt to minimize the distance between two probability distributions, we can use KL divergence loss. These loss functions will be introduced in later chapters as they are used.

The loss function is particularly important since it determines the goodness of the model fitness. Normally, when the loss is considered small enough in the training process, we can terminate training and get a trained model. Before getting into the detailed training process, we will first introduce how to update the neural network weights in the following section.

4.2.5 Back-propagation

The back-propagation algorithm is the core of deep learning since back-propagation calculating the gradient of the errors of all layers in a neural network is the prerequisite to update the weights of the neural network. Before explaining back-propagation, we will first introduce forward propagation.

Forward propagation

The aforementioned neural network calculations all belong to forward propagation (or forward pass). In the forward pass, input data are fed into the neural network for calculation and storage of intermediate variables (including outputs) in a forward order from the input layer to the output layer, as shown in Fig. 4.2.

For the sake of simplicity, let us assume that the input example is $\mathbf{x} \in \mathbb{R}^d$ and that our hidden layer does not include a bias term. The forward pass can be

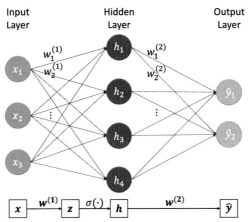

FIGURE 4.2 Forward propagation in a fully neural network, taking a two-layer neural network (MLP) as an example.

described as follows:

$$\mathbf{h} = \sigma(\mathbf{z}) = \sigma(\mathbf{W}^{(1)}\mathbf{x}), \tag{4.14}$$

where $\mathbf{W}^{(1)} \in \mathbb{R}^{h \times d}$ is the weight parameter of the hidden layer, i.e., the first layer, σ is the activation function, and $\mathbf{z} = \mathbf{W}^{(1)}\mathbf{x}$ is a virtual variable to show the calculation process. The obtained hidden variable $\mathbf{h}$ with the length of h will be input to the output layer, i.e., the second layer. Assuming the output layer does not have an activation function, the output variable of the neural network y can be obtained as the product of weight parameters of the output layer $\mathbf{W}^{(2)} \in \mathbb{R}^{q \times h}$ and the hidden variable $\mathbf{h}$:

$$\hat{\mathbf{y}} = \mathbf{W}^{(2)}\mathbf{h}. \tag{4.15}$$

As Fig. 4.2 shows, the output has two values $\hat{\mathbf{y}} = [\hat{y}_1, \hat{y}_2]$. Given the label $\mathbf{y}$, we can then calculate the loss term via the aforementioned loss function:

$$L = \text{loss}(\hat{\mathbf{y}}, \mathbf{y}). \tag{4.16}$$

Normally, the goal is to minimize the loss function in the neural network training process. To describe the whole computation process, we refer to J as the objective function, which is equivalent to the loss function $J = L$, in the following discussion,

Backward propagation

Back-propagation (BP) refers to the method of calculating the gradients of all the neural network parameters to update the network's weights. The forward computation process traverses from the input to the loss, while the back-propagation method traverses the network in reverse order, from the output to the input layer, according to the chain rule from calculus.

The chain rule can help calculate the gradients of all parameters in a neural network given the loss, as can be seen in the following example. Assume that we have functions $Y = f(X)$ and $Z = g(Y)$, in which the input and the output X, Y, Z are tensors of arbitrary shapes. By using the chain rule, we can compute the derivative of Z with respect to X via

$$\frac{\partial Z}{\partial X} = \text{prod}\left(\frac{\partial Z}{\partial Y}, \frac{\partial Y}{\partial X}\right). \tag{4.17}$$

Here, prod is the matrix–matrix multiplication operator.

To better illustrate the backward propagation of the neural network, we depict the backward computational graphs of the neural network just mentioned in Fig. 4.3 to show the dependencies of operators and variables within the calculation. The gray lines show the forward computation graph, and the red dotted lines describe the backward computation graph associated with the network. The weight parameters of this network are represented by $\mathbf{W}^{(1)}$ and $\mathbf{W}^{(2)}$. The BP process calculates the gradients of the objective J with respect to all variables and weight parameters in the neural network. Since $J = L$ and $\frac{\partial J}{\partial L} = 1$, the model's weight updating process targets to calculate $\partial L / \partial \mathbf{W}^{(1)}$ and $\partial L / \partial \mathbf{W}^{(2)}$. To accomplish this, we apply the chain rule and calculate the gradient of each intermediate variable and parameter, starting from L to the input X in the backward order.

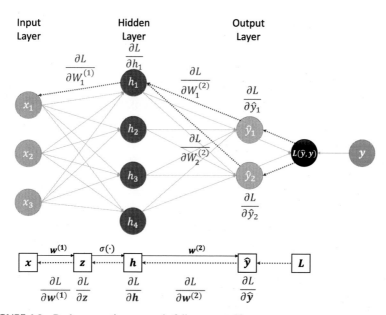

FIGURE 4.3 Back-propagation process in fully connected layers.

The first step is to compute the gradient of the objective function with respect to variables of the output layer according to the chain rule:

$$\frac{\partial J}{\partial \hat{y}_i} = \text{prod}\left(\frac{\partial J}{\partial L}, \frac{\partial L}{\partial \hat{y}_i}\right) = \frac{\partial L}{\partial \hat{y}_i}. \tag{4.18}$$

Based on Eq. (4.15), the gradient $\partial J / \partial W^{(2)}$ is calculated in a similar way:

$$\frac{\partial J}{\partial W_i^{(2)}} = \text{prod}\left(\frac{\partial J}{\partial \hat{y}_i}, \frac{\partial \hat{y}_i}{\partial W_i^{(2)}}\right). \tag{4.19}$$

To obtain the gradient with respect to $\mathbf{W}^{(1)}$, we need to continue back-propagation along the output layer to the hidden layer. We need to first calculate the gradient with respect to the hidden layer's outputs $\partial J / \partial \mathbf{h}$, which is given by

$$\frac{\partial J}{\partial \mathbf{h}} = \text{prod}\left(\frac{\partial J}{\partial \hat{\mathbf{y}}}, \frac{\partial \hat{\mathbf{y}}}{\partial \mathbf{h}}\right) = \mathbf{w}^{(2)} \frac{\partial L}{\partial \hat{\mathbf{y}}}. \tag{4.20}$$

Taking $\partial J / W_1^{(1)}$, as an example, the gradient is given by

$$\frac{\partial J}{\partial h_1} = \text{prod}\left(\frac{\partial J}{\partial \hat{y}_1}, \frac{\partial \hat{y}_1}{\partial h_1}\right) + \text{prod}\left(\frac{\partial J}{\partial \hat{y}_2}, \frac{\partial \hat{y}_2}{\partial h_1}\right) = w_1^{(2)} \frac{\partial L}{\partial \hat{y}_1} + w_2^{(2)} \frac{\partial L}{\partial \hat{y}_2}. \tag{4.21}$$

Since the activation function σ applies elementwise, calculating the gradient $\partial J / \partial \mathbf{z}$ of the intermediate variable $\mathbf{z}$ needs to use the element-wise multiplication operator, denoted by $\odot$:

$$\frac{\partial J}{\partial \mathbf{z}} = \text{prod}\left(\frac{\partial J}{\partial \mathbf{h}}, \frac{\partial \mathbf{h}}{\partial \mathbf{z}}\right) = \frac{\partial J}{\partial \mathbf{h}} \odot \sigma'(\mathbf{z}). \tag{4.22}$$

Finally, we can obtain the gradient $\partial J / \partial \mathbf{W}^{(1)}$ of the model parameters closest to the input layer. According to the chain rule, we get

$$\frac{\partial J}{\partial \mathbf{W}^{(1)}} = \text{prod}\left(\frac{\partial J}{\partial \mathbf{z}}, \frac{\partial \mathbf{z}}{\partial \mathbf{W}^{(1)}}\right) = \frac{\partial J}{\partial \mathbf{z}} \mathbf{x}^{\top}. \tag{4.23}$$

4.2.6 Validation dataset

In principle, we should not touch our test set until after we have chosen all our hyperparameters. Were we to use the test data in the model selection process, there is a risk that we might overfit the test data. Then, we would be in serious trouble. If we overfit our training data, there is always the evaluation on test data to keep us honest. But, if we overfit the test data, how would we ever know?

Thus, we should never rely on the test data for model selection. And yet we cannot rely solely on the training data for model selection either because we cannot estimate the generalization error on the very data that we use to train the model.

In practical applications, the picture gets muddier. While ideally we would only touch the test data once, to assess the very best model or to compare a small number of models to each other, real-world test data is seldom discarded after just one use. We can seldom afford a new test set for each round of experiments.

The common practice to address this problem is to split our data three ways, incorporating a validation dataset (or validation set) in addition to the training and test datasets. The result is a murky practice where the boundaries between validation and test data are worryingly ambiguous. Unless explicitly stated otherwise, in the experiments in this book, we are really working with what should rightly be called training data and validation data, with no true test sets. Therefore, the accuracy reported in each experiment of the book is really the validation accuracy and not a true test-set accuracy.

4.2.7 Underfitting or overfitting?

When we compare the training and validation errors, we want to be mindful of two common situations. First, we want to watch out for cases where our training error and validation error are both substantial but there is a little gap between them. If the model is unable to reduce the training error, that could mean that our model is too simple (i.e., insufficiently expressive) to capture the pattern that we are trying to model. Moreover, since the generalization gap between our training and validation errors is small, we have reason to believe that we could get away with a more complex model. This phenomenon is known as underfitting.

On the other hand, as we already discussed, we want to watch out for the cases when our training error is significantly lower than our validation error, indicating severe overfitting. Note that overfitting is not always a bad thing. Especially, with deep learning, it is well known that the best predictive models often perform far better on training data than on holdout data. Ultimately, we usually care more about the validation error than about the gap between the training and validation errors. Whether we overfit or underfit can depend both on the complexity of our model and the size of the available training datasets, two topics that we subsequently discuss.

4.3 Transportation applications

FNNs consisting of multiple linear layers can be considered as extended linear models. However, the activation functions bring nonlinear power to the FNN. Thus, an FNN can solve a big class of regression and logistic problems in the transportation domain.

4.3.1 Traffic prediction

In this subsection, we take the traffic prediction task as an example to show how FNNs can solve regression problems in the transportation domain. Traffic prediction is a broad topic in which roadway travel time, traffic speed, traffic volume, travel demand, origin-destination (OD), and mobility pattern can all be estimated or predicted using deep learning methods. These research areas indeed attract much attention in the past several years.

We first take the traffic volume prediction along a corridor as an example, the simplest prediction task is to predict the state at a specific location, such as a roadway link. Given a time series of traffic state $x = [x_1, x_2, ..., x_t]$, the prediction task is to take the series as input and generate a predicted state x_{t+1}. Hence, the traffic prediction problem can be formulated as learning a f function, such that $y = f(x)$, where y is the predicted value of x_{t+1}. Here, we can solve this prediction problem by using a multilayer fully connected neural network as the f function.

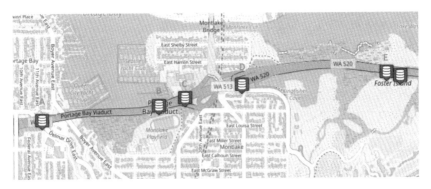

FIGURE 4.4 Traffic state collected by sensors along a corridor.

A more complicated case is to predict the traffic state along a corridor, as shown in Fig. 4.4, where the traffic states at different locations are all input into the model to predict the state at another place. For example, the task aims at predicting the traffic state at sensor **E,** taking the traffic states at sensor **A, B, C,** and **D** as input. If we use a basic FNN as the prediction model, the model structure is shown in Fig. 4.5. The model takes the traffic states at various locations x_A, x_B, x_C, x_D as input to generated a prediction output y_E. Since the prediction task needs to use historical data, the traffic state input collected by the i-th sensor is a time series of historical traffic states $x_i = [x_1^i, x_2^i, ..., x_t^i]$. Thus, the input of the model is $X = [x_A, x_B, x_C, x_D]$ is a two-dimensional (2D) matrix. In this case, this problem aims at fitting the f such that $y = f(X) = f([x_A, x_B, x_C, x_D])$.

These two examples focus only on the traffic prediction of a road segment or several consecutive segments. If all the segments of the road network are considered, a more comprehensive spatiotemporal data model should be adopted.

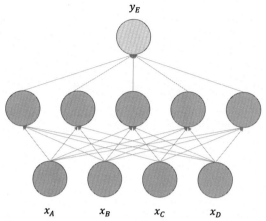

FIGURE 4.5 Traffic state prediction along a corridor, given states of sensor A, B, C, and D to predict the state of sensor E.

Although FNN is a candidate, more suitable deep learning models for solving the network-wide traffic prediction will be introduced in later chapters.

4.3.2 Traffic sign image classification

Traffic sign identification and classification is a necessary functional module of self-driving (autonomous) vehicles. It is a process of automatically recognizing traffic signs along the road, including speed limit signs, yield signs, merge signs, etc. It helps self-driving cars to properly parse and understand the roadway around them.

In this real-world, traffic sign recognition is normally divided into the two tasks, localization and recognition.

- Localization: Localize where the traffic sign is in an input image.
- Recognition: Take the localized regions of interest (ROI) of traffic signs, as shown in Fig. 4.6, and classify them accurately.

Though traffic sign recognition can be fulfilled in single forward-pass neural network-based object detectors, like Faster R-CNNs, Single Shot Detectors (SSDs), and RetinaNet, in this section, we take the traffic sign classification as an example to show how the fully connected neural network can solve a classification task in the transportation domain. The advanced computer vision-related deep learning methods will be introduced in detail in the convolutional neural network chapter.

The traffic sign classification problem, like all other classification problems, is a supervised learning problem, in which one chooses a model from the model set $\mathcal{F} = \{F_\theta\}$, where F_θ can be a neural network parameterized by θ, to classify a set of traffic sign images $\mathcal{X} = \{x_i\}$ as input to the correct labels $\mathcal{Y} = \{y_i\}$ as the output. If the model is a neural network-based classifier, consisting of

Some examples of training data

FIGURE 4.6   Traffic sign images for classification.

several fully connected layers, you can add a softmax layer as the last layer or directly use the cross-entropy loss function to ensure the trained model can output correct labels. Please note that the input data as an image is a 2D or 3D matrix. Using FNN to process the high-dimensional data requires a data flattening process that convert high-dimensional data into one-dimensional data. Thus, the traffic sign classification problem can still be formulated as learning a function f such that $y = f_\theta(x)$, where f can contain several FNN layers and other data preprocessing procedures.

4.4   Exercises

4.4.1   Questions

1. True or False questions:

- During forward propagation, in the forward function for a layer l, you need to know what is the activation function in a layer (sigmoid, tanh, ReLU, etc.). During back-propagation, the corresponding backward function also needs to know what is the activation function for layer l, since the gradient depends on it.
- The deeper layers of a neural network are typically computing more complex features of the input than the earlier layers.

2. Among the following, which ones are "hyperparameters"?
   a. size of the hidden layers;
   b. learning rate;
   c. number of iterations;
   d. number of layers in the neural network.

3. Which is NOT true about gradient descent?
   a. Gradient descent is intended to find the values of a function's parameters that minimize a cost function as much as possible.
   b. When maximizing an objective function, you can use gradient ascent instead of gradient descent.
   c. Gradient descent is a specific design method for neural network optimization.
   d. Stochastic gradient descent is a variant of the gradient descent method that is popular for neural networks training.

4. Consider a neural network with two hidden layers, each of which can be represented by $y_i = \sigma(x_{i-1} W_i + b_i)$. If the input of the neural network is $x_0 \in \mathbb{R}^{16 \times 1}$, the input of the second hidden layer is $x_1 \in \mathbb{R}^{32 \times 1}$, and the output of the second hidden layer is $y_2 \in \mathbb{R}^{16 \times 1}$, which of the following statement are true (check all that apply)?

   - W_1 will have shape (16, 16);
   - b_1 will have shape (16, 1);
   - y_1 will have shape (32, 1);
   - W_2 will have shape (32, 1);
   - b_2 will have shape (16, 1).

5. Why is zero initialization of weight not a good initialization technique?
6. Why data is normalization needed in most cases of deep learning models?
7. Why is ReLU the most commonly used activation function?
8. Explain the vanishing and exploding gradient problems.
9. What is the difference between epoch, batch, and iteration in neural networks?
10. List five scenarios that you want to use FNN to solve a transportation related problem.

Chapter 5

Convolution neural networks

5.1 Convolution neural network fundamentals

5.1.1 From fully connected layers to convolutions

A Convolution Neural Network (CNN) is a deep learning method that takes images or image-like inputs, assigns weights and biases to the components via a learning process, and carries out different tasks, such as differentiating one from the other, localizing interested objects, and learning the relationships between regions. A common question is why use CNNs over feed-forward neural networks? Images can certainly be flattened and fed into a feed-forward neural network as a normal feature vector, like the process shown in Fig. 5.1. This could be true when the image is simple, but a CNN is able to capture the spatial and temporal dependencies in multiple levels in an image when the input image is large and complicated.

There are two primary issues with fully connected layers. On the one hand, the number of network parameters is too large. For a image with the size of 400×300, it has 1.2×10^5 input dimensions. If the fully connected network has only one hidden layer with 1,000 neurons, the number of parameters to be learned is over 1.2×10^8. This will cost a lot of computation and storage resources and is far from the goal of efficient training and inference. On the other hand, the machine learning models we have discussed mainly deal with tabular data, which means the data is structured as rows corresponding to data points and columns corresponding to features. Even for image data, a 2-D image matrix is supposed to be converted to a 1-D vector before feeding the fully connected neural network. The models have assumed no mechanism for how the input data or features interact. This could be due to the simplicity of the data itself or a lack of knowledge in guiding the design of data feature interactions. In most of these cases, the fully connected neural network is probably the top choice since it can learn the high-dimensional features through the hidden nonlinear layers.

These issues with MLP motivate the use of CNNs. Imagine we have an image with a complicated scene in which our target is to locate a specific object (e.g., a dog, a person, a bike). What a human will do in this localization task is probably to move his/her line of sight from left to right along the top of the image, then from left to right in the middle of the image, and later slide the line of sight in the same way at the bottom of the image. In this way, the observer

Machine Learning for Transportation Research and Applications
https://doi.org/10.1016/B978-0-32-396126-4.00010-2

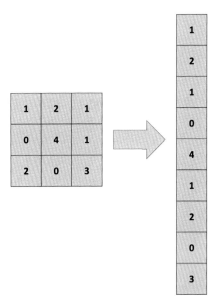

FIGURE 5.1   Flatten a 3 × 3 image into a 9 × 1 vector.

assigns each patch of the image a score by matching each patch to a target object image in his/her brain. In traditional computer vision-based object detection, a commonly adopted approach is similar to this intuitive way of searching by humans, called the Sliding Window approach (Glumov et al., 1995). In the sliding window approach, a learned classifier, often with hand-crafted features, assigns a score to each image patch in a sliding-window manner. Usually the sliding-window approach is applied on multiple scales of the input image to detect the same object in different scales. CNN is built upon this intuition, which results in much fewer parameters and at the same time makes use of the spatial relationships among pixels.

5.1.2   Convolutions

Convolution is the fundamental operation of a CNN. In this subsection, we briefly introduce the mathematical function *convolution*. The convolution is often between two functions f and g. The convolution operation is written as $f * g$. An asterisk usually indicates multiplication. In advanced calculus, the asterisk is used as a symbol for convolution. We will see that convolution is actually a type of multiplication. The formal definition of a convolution is Eq. (5.1):

$$(f * g)(t) = \int_{-\infty}^{\infty} f(z)g(t - z)dz. \tag{5.1}$$

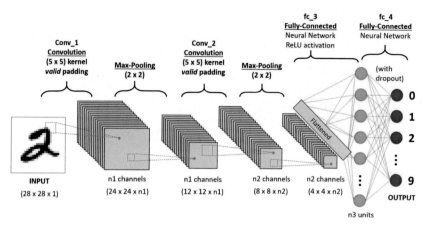

FIGURE 5.2   A typical CNN architecture for hand-written digits classification.

$g(t - z)$ indicates a flipped g function, and it is shifted by t. The integral turns into the sum for discrete objects, which is shown in Eq. (5.2):

$$(f * g)(k) = \sum_{a=-\infty}^{\infty} f(a)g(k - a). \qquad (5.2)$$

In the case of 2-D convolution for discrete objects, the operation is as follows:

$$(f * g)(k, l) = \sum_{a=-\infty}^{\infty} \sum_{b=-\infty}^{\infty} f(a, b)g(k - a, l - b). \qquad (5.3)$$

The g function is a 2-D function and flipped in both dimensions, and then shifted by k and l. The two-dimensional convolution of $f * g$ is essentially the sum of value multiplications along the two axes.

5.1.3   Architecture

A CNN architecture consists of convolution layers, pooling layers, activation functions, a flatten operator, a fully connected layer, and a classification function (e.g., Softmax). A typical CNN architecture is shown in Fig. 5.2. A convolution layer has multiple different kernels/filters that operate on the input image of the feature map from previous layers. Fig. 5.3 displays a convolution operation with one kernel/filter on a 2-D input image/feature. In case of RGB images or 3-D input features (e.g., videos), the kernel will have three dimensions instead of two. The kernel runs in a sliding-window manner on the input image. For example, the top left-hand element of the output convolved feature is obtained by the kernel's operation on the top left-hand 3×3 grid on the input image, i.e., $-2 = 1 \times 2 + 0 \times 1 + 1 \times 0 + 0 \times 0 + 0 \times 1 + 0 \times 0 + (-1) \times 1 + 0 \times 0 + (-1) \times 3$.

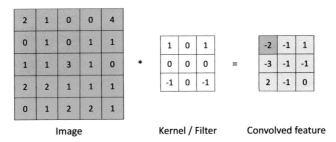

FIGURE 5.3 Convolution operation.

A concept in the convolution operation is *stride*, which determines the step a kernel moves in the sliding window operation. If *stride length* = 1, a kernel will operate on all the windows of its size in the input image. The case shown in Fig. 5.3 is an example of stride length = 1. If *stride length* = 2, a kernel will skip one image pixel to run the next convolution operation. Another critical component of a convolution layer is the activation function. Similar to a feed-forward neural net, sigmoid and ReLU are the two commonly used activation functions. However, ReLU is more dominant than Sigmoid in CNN architectures because of its advantages computational efficiency and in having no vanishing gradient. The output size of a convolution layer is

$$S_{output} = \frac{S_{input} - S_{filter}}{S_{stride}} + 1, \tag{5.4}$$

where S_{output}, S_{input}, S_{filter}, S_{stride} are the sizes of the output, input, filter, and stride of a convolution layer. Also, the number of filters becomes the number of channels in the output feature map of a convolution layer.

A convolution layer and a pooling layer together form a complete layer of a CNN. A pooling layer often appears after a convolution layer. It reduces the size of the convolved feature to further improve the processing efficiency. A pooling operation extracts dominant features in a small window and, at the same time, is a positional and rotational invariant. There are two commonly used pooling operations: max pooling and average pooling. Max pooling extracts the maximum value of a region in the output convolved feature map from the convolution layer; average pooling returns the average value of the same region. An example is shown in Fig. 5.4. Max pooling is less sensitive to noises, while average pooling is influenced by all values in the pooling region. Therefore, max pooling in general performs better than average pooling. The last few layers are usually fully connected layers, which have the same structure as the feed-forward neural network structure introduced in last chapter.

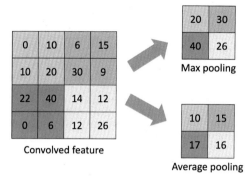

FIGURE 5.4   Max pooling and average pooling operations.

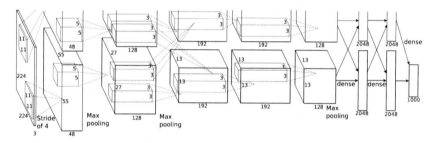

FIGURE 5.5   The AlexNet CNN architecture.

5.1.4 AlexNet

AlexNet was the very first CNN architecture that used GPU to boost performance (Krizhevsky et al., 2012). It was proposed and won the Imagenet Visual Recognition Challenge in 2012. The architecture is presented in Fig. 5.5. AlexNet has eight layers, among which five are convolution layers and three are fully connected layers. They all use ReLu as the activation function. Max pooling is the pooling operation that follows all convolution layers. The input to AlexNet is $227 \times 227 \times 3$. The first convolution layer has 96 filters of size 11×11 with stride 4, and the output is $55 \times 55 \times 96$. The first max-pooling layer is of size 3×3 and stride 2, resulting in a feature map of $27 \times 27 \times 96$. The second convolution layer has 256 filter of size 5×5, and with stride 1 and padding 2, resulting in a feature map of size $27 \times 27 \times 256$. The second max-pooling layer is 3×3 with stride 2, and the output is of size $13 \times 13 \times 256$. Likewise, AlexNet has third, fourth, and fifth convolution layers with max pooling and ReLu operations. The fully connected layers have 4,096, 4,096, and 1,000 neurons, respectively. Softmax function is the activation function for the last fully connected layer. There are two dropout layers following the first two fully connected layers. The dropout rate is 0.5. AlexNet in total has about 62.3 million parameters in its architecture.

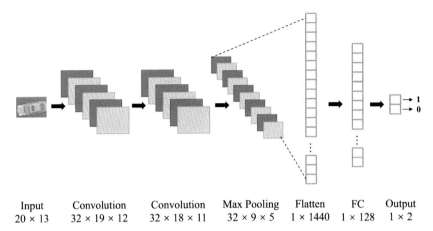

Input	Convolution	Convolution	Max Pooling	Flatten	FC	Output
20×13	$32 \times 19 \times 12$	$32 \times 18 \times 11$	$32 \times 9 \times 5$	1×1440	1×128	1×2

FIGURE 5.6   The proposed CNN architecture.

5.2 Case study: traffic video sensing

We showcase an example in traffic video sensing using CNN. More details of this example can be found in the authors' paper Ke et al. (2018a). Specifically, we use CNN for vehicle classification in unmanned aerial vehicle (UAV) images. With a trial and error process, the architecture of our CNN (see Fig. 5.6) was chosen to contain two convolution layers, one pooling layer, and one hidden FC layer. The two convolution layers have a same dimension of $32 \times 2 \times 2$ with a sigmoid activation function. The pooling layer is added to downsample the second convolution layer's outputs by a scale factor of 2. And the FC layer with 128 nodes is added between the pooling layer and the final outputs. Compared to other popular CNN architectures, such as AlexNet (Krizhevsky et al., 2012) and VGG (Simonyan and Zisserman, 2014), this CNN structure is lightweight with many fewer layers and parameters.

This is motivated by the requirement for real-time operation on UAVs and a smaller number of categories (i.e., vehicle and background). It has been found that two convolution layers can already satisfy the accuracy requirement. It is worth noting that there is no pooling layer in between the two convolution layers. This is because the training and testing losses turn out to be higher, while the overall detection speed is not significantly improved if adding the pooling layer. Based on our tests and analyses, the increased losses are mainly caused by the small image size after pooling. Since the dimension of our image samples is 20×13, if they are downsampled by the pooling layer, the features extracted by the second convolutional layer would not be as representative.

We present in Fig. 5.8 two detection results from a classifier with traditional hand-crafted features (Haar cascades) and the proposed CNN classifier. It can be seen that the false detection rate of the traditional classifier (top) is rather high and the detection of CNN looks neat. However, the processing time of

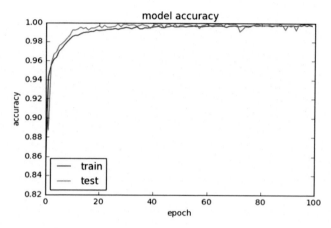

FIGURE 5.7 The training and testing accuracy curves.

CNN is much longer than with haar cascades, even with the lightweight CNN structure design. Therefore, instead of directly applying CNN to a UAV image with a sliding-window detection manner, a combined detector Haar + CNN is a more optimal solution in this case. The Haar classifier detects vehicles first in the image, and then the CNN just examines the remaining candidates (the red bounding boxes in the top image of Fig. 5.8). This method results in high accuracy and efficiency.

The training of CNN was done on 18,000 samples and testing on 2,000 samples. RMSprop (Root Mean Square Propagation) was selected as the optimizer because of its better performance than others like traditional SGD (Stochastic Gradient Descent) in similar cases based on experience and tests. The batch size for optimization was set to 30. Our CNN vehicle classifier reached 99.55% classification accuracy on the test data in 100 epochs of training, which was very encouraging. The model accuracy curves (training and testing) is shown in Fig. 5.7.

5.3 Case study: spatiotemporal traffic pattern learning

In this subsection, we introduce how CNN can be applied to spatiotemporal traffic pattern learning in traffic speed prediction. We introduce a two-stream, multichannel convolution neural network (TM-CNN) for multilane traffic speed prediction with the consideration of traffic volume impact. More details of the example can be found in our paper Ke et al. (2020c). In the proposed model, we develop a data conversion method to convert both the multilane speed data and multilane volume data into multichannel spatiotemporal matrices. We design a CNN architecture with two streams, where one takes the multichannel speed matrix as input and another takes the multichannel volume matrix as input. A fusion method is further implemented for the two streams. Specifically, convolution

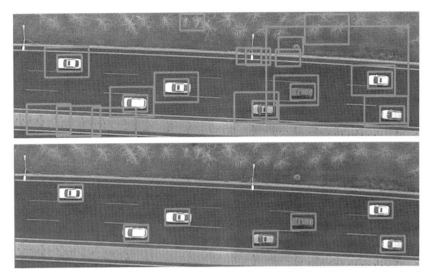

FIGURE 5.8 The vehicle detection results from a hand-crafted classifier Haar cascade (top) and the proposed CNN (bottom).

layers learn the two matrices to capture traffic flow features in three dimensions: the spatial dimension, the temporal dimension, and the lane dimension. Then, the output tensors of the two streams will be flattened and concatenated into one speed–volume vector, and this vector will be learned by the FC layers. Accordingly, a new loss function is devised considering the volume impact in the speed prediction task.

The first step of our methodology is modeling the multilane traffic flow as multichannel matrices. We propose a data conversion method to convert the raw data into spatiotemporal multichannel matrices, in which traffic on every individual lane is added to the matrices as a separate channel. This modeling idea comes from CNN's superiority to capture features in multichannel RGB images. In RGB images, each color channel has correlations yet differences with the other two. This is similar to traffic flows on different lanes where correlations and differences both exist. Thus, averaging traffic flow parameters at a certain milepost and time stamp is like doing a weighted average of the RGB values to get the grayscale value. In this sense, previous methods for traffic speed prediction are designed for "grayscale images" (spatiotemporal prediction for averaged speed) or even just a single image column (speed prediction for an individual location). In this study, the proposed model manages to handle lane-level traffic information by formulating the data inputs as "RGB images."

This data conversion method diagram is shown in Fig. 5.9. There are loop detectors installed at k different mileposts along this segment, and the past n time steps are considered in the prediction task. We denote the number of lanes as c. Without loss of generality, it is assumed that the number of lanes is three

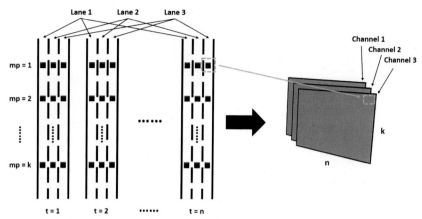

FIGURE 5.9 The data input modeling process of converting the multilane traffic flow raw data to the multichannel spatiotemporal matrix.

in Fig. 5.9 for the sake of illustration. Single-lane traffic would be represented by two $k \times n$ spatiotemporal 2-D matrices, where one is for speed and another for volume. We denote them as I_u for speed and I_q for volume. We define the speed value and volume value to be u_{ilt} and q_{ilt} respectively for a detector at milepost i ($i = 1, 2, \ldots, k$) and lane l ($l = 1, 2, \ldots, c$) at time t ($t = 1, 2, \ldots, n$). Note that each u_{ilt} or q_{ilt} is normalized to between 0 and 1 using min-max normalization since the speed and volume have different value ranges. Hence, in the speed and volume matrices with the size $k \times n \times c$, we construct the 12 matrices using Eqs. (5.5) and (5.6)

$$I_u(i, t) = (u_{i1t}, u_{i2t}, \ldots, u_{ict}) \tag{5.5}$$
$$I_q(i, t) = (q_{i1t}, q_{i2t}, \ldots, q_{ict}), \tag{5.6}$$

where i and t are the row index and column index of a spatiotemporal matrix, representing the milepost and the time stamp, respectively. $I_u(i, t)$ and $I_q(i, t)$ denote the multichannel pixel values of the speed and the volume. The number of channels corresponds to the number of lanes c. Each element in the 2-D multichannel matrices is a c-unit vector representing c lanes' traffic speeds or volumes at a given milepost i and time t. In the three-lane example in Fig. 5.9, the spatiotemporal matrices have three channels. Mathematically, the spatiotemporal multichannel matrices for traffic speed (X_u) and volume (X_q) can be denoted as Eqs. (5.7) and (5.8):

$$X_u = \begin{bmatrix} I_u(1, 1) & I_u(1, 2) & \ldots & I_u(1, n) \\ I_u(2, 1) & I_u(2, 2) & \ldots & I_u(2, n) \\ \vdots & \vdots & & \vdots \\ I_u(k, 1) & I_u(k, 2) & \ldots & I_u(k, n) \end{bmatrix} \tag{5.7}$$

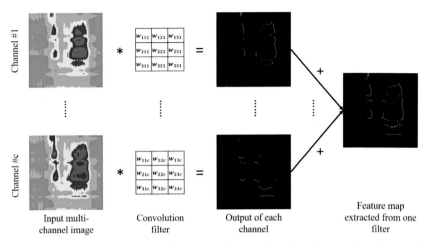

FIGURE 5.10 The convolution operation to extract features from the multichannel spatiotemporal traffic flow matrices.

$$X_q = \begin{bmatrix} I_q(1,1) & I_q(1,2) & \cdots & I_q(1,n) \\ I_q(2,1) & I_q(2,2) & \cdots & I_q(2,n) \\ \vdots & \vdots & & \vdots \\ I_q(k,1) & I_q(k,2) & \cdots & I_q(k,n) \end{bmatrix}. \tag{5.8}$$

With the reorganized input as a multichannel matrix X (X could be X_u or X_q), the basic unit of a convolution operation is shown in Fig. 5.10. On the far left of the figure, it is the input spatiotemporal matrix or image X. Every channel of the input matrix is a 2-D spatiotemporal matrix representing the traffic flow pattern on the corresponding lane. On the top of the left-hand column, channel 1 displays the traffic pattern of lane 1; and on the bottom, the pattern of lane c is presented. The symbol "*" denotes the convolution operation in Fig. 5.10. Since our input is a multichannel image, the convolution filters are also multichannel. In the figure, a $3 \times 3 \times c$ filter is drawn, while the size of the filter can be chosen as needed. The values inside the cells of a filter are the weights of the CNN, which are automatically modified during the training process. The final weights are able to extract the most salient features in the multichannel image. The convolution operation outputs a feature map for each channel, and they are summed up to be the extracted feature map of this convolution filter in the current convolution layer. With multiple filters applied to the same input image, a multichannel feature map will be constructed, which serves as the input to the next layer.

In order to learn the multilane traffic flow patterns and predict traffic speeds, a CNN structure is designed (see Fig. 5.11). Compared to a standard CNN, the proposed CNN architecture is modified in the 11 following aspects: (1) The network inputs are different, that is, the input image is a spatiotemporal image built

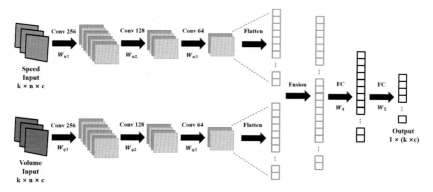

FIGURE 5.11 The proposed TM–CNN architecture.

by traffic sensor data, and it has multiple channels that represent the lanes of a corridor. Moreover, the pixels values' range is different from a normal image. For a normal image, it is 0–255; however, here it ranges 0–15 to either the highest speed (often the speed limit) or the highest volume (often the capacity). (2) The neural network has two streams of convolution layers, which are for processing the speeds and volumes, but most CNNs have only one stream of convolution layers. The purpose of having two streams of convolution layers is to integrate both speed information and volume information into the model so that the network can learn the traffic patterns better than only learning speed. To combine the two streams, a fusion operation that flattens and concatenates the outputs of the two streams are implemented between the convolution layers and the FC layers. The fusion operation concatenation is chosen instead of addition or multiplying because the concatenation operation is more flexible for us to modify each stream's structures. In other words, the concatenation fusion method allows the two streams of convolution layers to have different structures. (3) The extracted features have unique meanings and are different from image classifications or most other tasks. The extracted features here are relationships among road segments, time series, adjacent lanes, and between traffic flow speeds and volumes. (4) The output is different, i.e., our output is a vector of traffic speeds of multiple locations at the future time rather than a single category label or some bounding-boxes' coordinates. The output itself is part of the input for another prediction, while this is not the case for most other CNN's. (5) Different from most CNN's, the proposed CNN does not have a pooling layer. The main reason for not inserting pooling layers in between convolution layers is that our input images are much smaller than regular images for image classification or object detection. Regular input images to a CNN usually have hundreds of columns and rows, while the spatiotemporal images for roadway traffic are not that large. In this research and many existing traffic prediction studies, the time resolution of the data is five minutes, which means that, even when using two-hour data for prediction, there are only 24 time steps. Thus, we do not risk losing information by pooling. (6) The loss function is devised to contain both speed and volume information. For traditional image classification CNN's, the loss function is the cross-entropy loss. And for traffic speed prediction tasks, the loss function is commonly the Mean Squared Error (MSE) function with only speed values. However, in this research, we add a new term in the loss function to incorporate the volume information. We denote the ground truth speed vector and volume vector as Y_u and Y_q, and the predicted speed vector and volume vector as $\hat{Y}_u$ and $\hat{Y}_q$. Note that Y_u, Y_q, $\hat{Y}_u$, and $\hat{Y}_q$ are all normalized between 0 and 1. The loss function L is defined in Eq. (5.9) by summing up the MSEs of speed and volume. The volume term $\lambda \| \hat{Y}_q - Y_q \|_2^2$ is added to the loss function to reduce the probability of overfitting by helping the model better understand the essential traffic patterns. This design improves the speed prediction accuracy on the test dataset with proper settings of λ. Our suggested value of λ is between 0 and 1, considering that the volume term that deals with overfitting should still have a lower impact than the speed term on speed prediction problems.

$$L = \left\| \hat{Y}_u - Y_u \right\|_2^2 + \lambda \left\| \hat{Y}_q - Y_q \right\|_2^2. \tag{5.9}$$

In the proposed TM–CNN, the inputs are our multichannel matrices X_u and X_q with the dimension of $k \times n \times c$. The filter size is all $2 \times 2 \times c$ in order to better capture the correlations between each pair of adjacent loops, as well as adjacent times. The number of filters for each convolution layer is chosen based on experience and consideration of balance efficiency and accuracy. The last convolution layer in each of the two streams is flattened and connected to a FC layer. This FC layer is fully connected with the output layer as well. The length of the output vector $\hat{Y}_u$ is $1 \times (k \times c)$ since the prediction is for one future step. All activations except the output layer use the ReLU function. The output layer has a linear activation function, which is adopted for regression tasks. Eqs. (5.10) and (5.11) describe the derivations mathematically from inputs to the outputs of the last convolution layers,

$$\hat{Y}_u^{conv} = \varphi \left\{ W_{u3} * \varphi \left[W_{u2} * \varphi \left(W_{u1} * X_u + b_{u1} \right) + b_{u2} \right] + b_{u3} \right\} \tag{5.10}$$

$$\hat{Y}_q^{conv} = \varphi \left\{ W_{q3} * \varphi \left[W_{q2} * \varphi \left(W_{q1} * X_q + b_{q1} \right) + b_{q2} \right] + b_{q3} \right\}, \tag{5.11}$$

where $\hat{Y}_u^{conv}$ and $\hat{Y}_q^{conv}$ are the intermediate speed and volume outputs of the CNN in between the last convolution layers and the flatten layers, W_{ui} and W_{qi} ($i = 1, 2, 3$) are the weights for the convolutions, b_{ui} and b_{qi} ($i = 1, 2, 3$) are the biases, and $\varphi(\cdot)$ is the ReLU activation function. After obtaining these two intermediate outputs, we flatten them and fuse them into one vector, and then further learn the relationships between the volume feature map and the speed feature map using FC layers.

As aforementioned, we choose concatenation as the fusion function for the two flattened intermediate outputs to allow the customization of variou neural network designs of the two streams. Customized streams could result in two intermediate outputs of different dimensions. While concatenation would still successfully fuse the two outputs together and support the learning of speed–volume relationships by the FC layers, most other fusion operations require the two vectors to have the same length. This fusion process is mathematically represented in Eq. (5.12) as follows:

$$\hat{Y}_u, \hat{Y}_q = W_5 \times \varphi \left[W_4 \times \text{Conv} \left(F \left(\hat{Y}_u^{\text{conv}} \right), F \left(\hat{Y}_q^{\text{conv}} \right) \right) + b_4 \right] + b_5, \quad (5.12)$$

where W_4 and W_5 are the weights for the two FC layers, b_4 and b_5 are the biases, $F(\cdot)$ is the flatten function, and $\text{Conc}(\cdot)$ is the concatenation function.

5.4 Case study: CNNs for data imputation

This study puts forward an innovative method for missing traffic data imputation that solves the imputation problem from the perspective of the spatiotemporal 2D image using a deep learning-based image inpainting approach. Traditionally, the traffic data imputation approach applied the pure mathematical methods, which cannot make full use of both spatial and temporal information. The basic idea of our approach is first transforming raw data into a 2-D image and then applying a CNN based-image inpainting approach to impute the missing data. This study focuses on imputing traffic volume data. Therefore, the raw data is traffic volume data. This method has good scalability, which means it can apply to other types of data, such as traffic velocity data, with little modification.

The first step of data imputation is transforming the raw traffic volume data into 2-D images. The horizontal axis (or x-axis) denotes the loop detectors' IDs which are selected continuously from one road or several intersecting roads. In this way, the selection of loop detectors ensures all loop data is spatially related because two adjacent columns are spatially neighboring. The vertical axis (or y-axis) denotes the detection time stamp. Thus, the image size is $N \times M$, where N represents the loop detector number and M represents the time-stamp number. The value of each pixel represents the traffic volume obtained from loop detectors during a specific time period. When displaying the traffic volume in an image, measurements are linearly mapped into the range 0–255 in grey scale. Thus, the traffic volume can be intuitively observed in both the spatial and temporal dimensions. An example 2-D traffic volume image is shown in Fig. 5.12(a).

When part of the data is missing, the corresponding pixel value is set to -1 in the spatiotemporal matrix, and the corresponding missing pixel is mapped into 255 in the data volume image (pure white color). The dark parts in Fig. 5.12 indicate the existence of traffic volume data. The darker is the pixel is, the lower is the traffic volume. As mentioned, white pixels indicate a lack of data at such a spatiotemporal location (i.e., a location where data imputation is needed). There are three types of missing data in an image: the central region case, the random

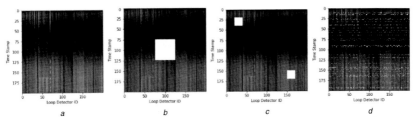

FIGURE 5.12 Traffic volume image with different missing status: (a) complete data without any missing region, (b) cata with missing central regions, (c) cata with randomly missing blocks, (d) data with randomly missing regions.

block case, and the random region case. Cases involving the central region indicate that the missing part is one entire rectangular region located in the center of the image. This means that several successive loop detectors lost data during the same long period. The central region case is shown in Fig. 5.12(b). The random block case occurs when there are several small missing rectangular regions randomly located in the image as shown in Fig. 5.12(c). The random region case, which is shown in Fig. 5.12(d), occurs when the shape and location of the missing regions are both random. The last two situations represent cases for which there are several loop detectors losing data randomly over a short period. The difference between the last two cases is the length of the missing period.

5.4.1 CNN-based imputation approach

The CNN is widely used in the object classification and recognition tasks. It can extract high-level features that can filter some disturbances such as light and position. Thus, applying a CNN to analyze traffic volume image can obtain inner correlations among different loop detectors and time stamps.

The basic idea is combining features of the encoder–decoder pipeline and the loss function of the generative adversarial network (GAN). The main architecture is an encoder–decoder pipeline. The encoder at first transforms the input image with missing parts into the feature map. Then, the decoder restores the feature map back to a complete image. The inpainting regions can thus be extracted by putting the mask of missing parts on the output image. The encoder and decoder are connected by a fully connected layer. The structure of the CNN-based imputation approach is shown in Fig. 5.13.

The encoder adopts the AlexNet architecture (Krizhevsky et al., 2012) which was proposed in 2012 and became one of the most popular architectures. CNN comes from regular neural networks that receive input and transform it through a series of hidden layers. Each hidden layer is made up of a set of neurons, where each neuron is fully connected to all the neurons in the previous layer. The neurons in a single-layer function are completely independent and do not share any connections. The last fully connected layer is also called the 'output layer'. When this network is used for classifications, it represents the class

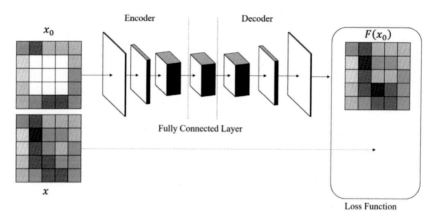

FIGURE 5.13   CNN architecture for data imputation.

scores. Compared to regular neural network, a CNN takes advantage of the fact that the input is an image and arranges neurons in three dimensions: width, height, and depth. To be more specific, the hidden layers in CNN typically consist of convolution layers, pooling layers, and fully connected layers in order, instead of only fully connected layers in regular neural networks. In this case, the input image size is 200×200, and the output feature map is $6 \times 6 \times 180$.

The convolution layer computes the output of neurons that are connected to local regions in the input volume. After the convolution layer, the pooling layer will perform a downsampling operation in the spatial dimensions. There are three convolution layers, and each one is followed by one pooling layer. To make the following explanation clear, one convolution layer and one pooling layer are referred to as a ConvoPooling layer. There are three ConvoPooling layers in total. Various layer sizes are tested, and the following combination is the optimized one. For the first layer, there are 20 filters with size 9×9. The pooling size is 4×4. Thus, the output dimension is $48 \times 48 \times 20$. The filter size for the second layer is 9×9, and there are 60 of them. The pooling size is 2×2. The output dimension is $20 \times 20 \times 60$. The last convolution layer has 180 filters with size 9×9, and the pooling size is 2×2. The output dimension of $6 \times 6 \times 180$.

The fully connected layer intends to spread information with activations of each feature map. The input layer has 180 feature maps of size 6×6, and it will output 180 feature maps of dimension 6×6. In contrast to other fully connected layers, it has no parameter connecting different feature maps and only spreads information within feature maps. Thus, the number of parameters of this layer is $180 \times 6 \times 6$ rather than $180^2 \times 6 \times 6$, which is the dimension in the other fully connected layers.

The rest of the pipeline is the decoder following the fully connected layer. The core of decoder has three up-convolution layers with the rectified linear unit (ReLU) activation function. The definition of ReLU is $f(x) = \max(0, x)$.

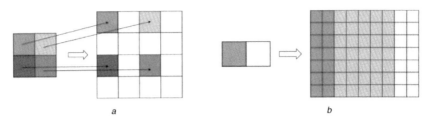

FIGURE 5.14 Principle/setup of decoder (a) 2 × 2 Unpooling, (b) 2 × 2 Unpooling and 5 × 5 convolution.

In contrast to traditional convolution operation, the up-convolution is still a convolution but produces a higher resolution image, which can be learned as a convolution following the unpooling (i.e., increasing the spatial span). The unpooling is just replacing each pixel of the feature map by a $m \times m$ block with the pixel value in the top left-hand corner. The rest is set to 0. The basic principle of the decoder is shown in Fig. 5.14. The different color means the different pixel values in the image. The orange part (mid gray in print version) in Fig. 5.14(b) is the weighted sum of red (dark gray in print version) and yellow (light gray in print version) parts in the convolution operation.

The context encoder–decoder pipeline is trained by regressing the content with missing regions to the originally complete content. The only difference between these two kinds of contests is the missing region, while other areas are the same. Thus, the loss function is important for optimizing the pipeline. The loss function selected in this paper is jointly comprised of two functions: reconstruction (L2) loss and adversarial loss. The reconstruction loss will capture the overall structure of the missing region and correlations about its context. It will average the influence of multiple modes in predictions. The adversarial loss tends to choose a particular mode from the distribution to make the prediction look realistic.

For the input image x_0 with missing regions, the context encoder–decoder pipeline F produces an output $F(x_0)$. The actual input image without missing regions is defined as x. The joint loss function computes only the loss of x and $F(x_0)$ in the missing regions. To explain the loss function clearly, the missing region mask $\hat{M}$ is defined. The corresponding values of the missing region are set to 1, while others are set to 0. The reconstruction loss function is derived from the Euclidean distance which is shown in (1) as follows:

$$L_{\text{rec}}(x) = \hat{M} \odot (x - F(x_0))_{\frac{2}{2}}^{2}, \qquad (5.13)$$

where $\odot$ is a convolution operation for a specific when combining with mask $\hat{M}$. After applying this operation, the loss function effect range is limited in the missing region. This loss function makes the pipeline tend to produce the rough outline of the inpainting image.

The adversarial loss function comes from GAN. One important parameter of GAN, called D, is the adversarial discriminator that takes both the prediction result and originally complete object into consideration, and tries to distinguish between them. The adversarial loss function is shown in the following equation:

$$L_{\text{adv}}(x) = \max_{D} E_{x \in \chi} \left[\log(D(x)) - \log(1 - D(F(x_0))) \right], \qquad (5.14)$$

where both F and D are optimized by stochastic gradient descent. This loss function reduces the imputation errors and makes the output of the pipeline closer to the actual data. Therefore, the joint loss function is shown in (3) as follows. The joint loss is the weighted sum of both reconstruction loss and adversarial loss. The loss function is important for the final accuracy. The optimized weights are $\lambda_{\text{rec}} = 0.35$ and $\lambda_{\text{adv}} = 0.65$

$$L = \lambda_{\text{rec}} L_{\text{rec}} + \lambda_{\text{adv}} L_{\text{adv}}. \qquad (5.15)$$

5.4.2 Experiment

The dataset used in this study was collected from loop detectors on the Interstate 5 (I-5) freeway in Washington State. In total, three months of data from January 2016 to March 2016 were collected from 256 loop detectors and used for the experiment in our study. With records collected at a 5-min time interval, this generates more than 50,000 records, which is considered a sufficient sample size for the case study.

Based on the literature review, the missing ratio is normally between 10 and 15%, but the ratio could be as high as 90% in some extreme cases. This paper takes the whole traffic volume dataset as the raw data with no missing information. Manual operations have been performed on the raw data to remove part of it with certain ratios ranging from 5 to 50% (5% gain). This is better than using raw data with originally missing parts because the comparison between estimated values and ground truth data can be performed. The dimension of one training data image is set as 200×200, which means there are 200 detectors in the spatial dimension and 200 records in the temporal dimension. It is possible to make the images with other dimensions, such as 100×100. To avoid a lack of features, the dimension should not be small. The training process and testing process may take a longer time when the dimension is too large.

The sample size of the training dataset is 14,000. There are ten groups of test data images, each of which has the same dimensions as the training images, but with different missing ratios. Each group has ten images with the same missing ratio, and the error of each group is calculated by averaging the estimated errors of all images in the group. Besides the proposed CNN-based approach, another two baseline approaches are implemented for comparative purposes: a DSAE-based approach and a BPCA approach. The comparison results highlight the strengths of our approach.

The evaluations of the imputation approaches are measured by the error of the imputed data. Three widely used criteria are adopted in this paper that focus on different perspectives of error. The first criterion considered is the root mean square error (RMSE), which is shown in Eq. (5.16). It represents the sample standard deviation of the differences between predicted values and observed values. Further, it just compares forecasting errors of different models for particular data, but not the whole dataset because it is scale dependent:

$$\text{RMSE} = \sqrt{\frac{1}{n} \sum_{i=1}^{n} \left(x_i^r - y_i\right)^2}. \tag{5.16}$$

The second criterion used is the mean absolute error (MAE) shown in Eq. (5.17), and it is the sum of two components—quantity and allocation disagreement. The quantity disagreement is defined as the absolute value of the mean error, and the allocation disagreement is the negative quantity disagreement. The MAE calculates the average absolute difference:

$$\text{MAE} = \frac{1}{n} \sum_{i=1}^{n} \left|x_i^r - y_i\right|. \tag{5.17}$$

The last evaluation criterion used is the mean percentage error (MPE) shown in Eq. (5.18). It computes the average of absolute percentage errors by which predictions of a model differ from actual values of the quantity being predicted:

$$\text{MPE} = \frac{1}{n} \sum_{i=1}^{n} \frac{\left|x_i^r - y_i\right|}{x_i^r}, \tag{5.18}$$

where n is the total number of the missing data, y_i is the ith imputed data, and x_i^r is the y_i's corresponding raw data.

This model is built on TensorFlow via Python. Thus, the training and testing work are both done on the TensorFlow framework. The training data is divided into 200 batches with each batch containing 70 images before being put into the network for the training process.

The RMSE, MAE, and MPE curves of the three approaches are shown in Fig. 5.15. In order to reduce the influence of the various selections of missing regions, the mean operation is based on ten evaluation results for the same method and the same missing ratio. However, the missing regions are randomly chosen for each evaluation. According to Fig. 5.15, the overall trends of the three curves are similar. When the missing ratio is < 30%, it is difficult to tell which approach is the best since they have similar errors. Also, it appears that random disturbances can greatly influence the final results. In our experiment, the BPCA has a better performance than the DSAE. This is because BPCA mainly focuses on previous information at the same detector and will not be affected by

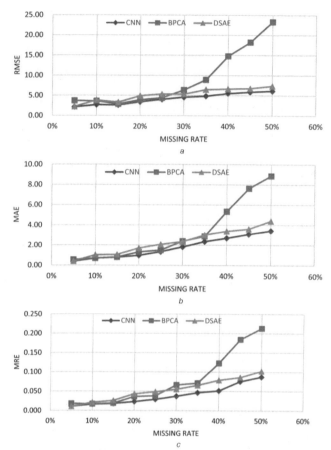

FIGURE 5.15   Error comparison with different criterion: (a) RMSE result, (b) MAE result, (c) MPE result.

nearby detector data. Additionally, the Bayesian model can correct errors on a general level. When the missing ratio is low, the temporal information plays a major role in imputation. The spatial information plays a minor role and might even become a disturbance. Although the proposed approach also makes use of spatial information, it can also extract higher-level features than the DSAE, which means that it can filter out portions of useless information. As the missing ratio increases, the error of the BPCA increases much faster than the other two approaches. The temporal information itself is not sufficient for data imputation when there is a lot of data missing. Therefore, the DSAE and the CNN start to show the strength of their data imputation approaches due to the fact that they combine both temporal and spatial information. Ultimately, the CNN still performs better than the DSAE in almost all cases.

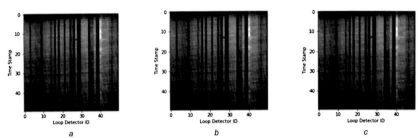

FIGURE 5.16 Imputation results with CNN-based approach: (a) Actual traffic volume data, (b) Imputation traffic volume data with 10% missing ratio, (c) Imputation traffic volume data with 40% missing ratio.

The imputation results using the CNN-based approach are shown in Fig. 5.16 with three missing ratios of 0, 10, and 40%. To make these figures easy for configuration, the loop detector number is from 0 to 49 and the time stamp number of 50 on 20 January 2016. The x-axis indicates the loop detector number corresponding to the spatial information and the y-axis indicates the time stamp. These figures use different colors to show the traffic volume, where the bright colour means that the traffic volume is high, and the dark colour means that the traffic volume is low. The three images in Fig. 5.16 are very similar in appearance, which indicates that our approach imputes the data well on a general level.

To see the detail of the imputation results, the data of loop detector No. 50 with time stamp number of 100 is extracted from one column of the preceding imputation results. Fig. 5.17 shows the imputation results when the missing ratio is 10 and 40%. The blue line (dark gray in print version) represents the imputed data points, and the orange line (mid gray in print version) represents the original data points. Similarly, the imputation results applying the DSAE and BPCA approaches are shown in Figs. 5.18 and 5.19. Hence, the x-axis shows the time stamp whose interval is 5 min. The y-axis shows the traffic volume and its unit is vehicles (observed in a 5 min interval). The imputation results of the CNN based approach show that it can impute the missing portion of the data well and recover it in a manner such that values are close to the actual ones observed in the ground truth data. Further, the distribution of errors on each missing point is stable, and there are no outliers.

Similar to the CNN-based approach, the DSAE approach also shows good stability in error distribution (Fig. 5.18). However, the average and variance of the error are larger than those for our approach, partly because the DSAE approach can only extract low-level features. The texture of the traffic volume image has a strong pattern which means the texture will change regularly. Therefore, the DSAE is more likely to be influenced by noise, and its error will be greater than those for the CNN-based approach on average.

The BPCA shows greater error value earlier in the time series because the method mainly depends on previous data (Fig. 5.19). In the beginning, the

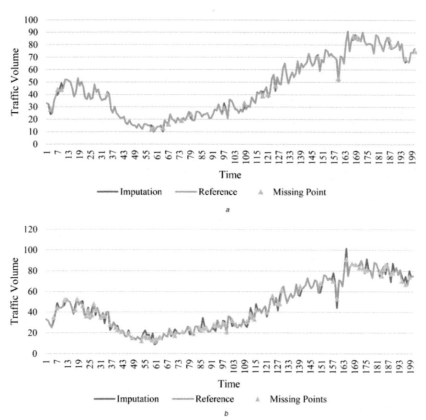

FIGURE 5.17 Imputation results with CNN-based approach: (a) 10% Missing ratio, (b) 40% Missing ratio.

amount of previous data is small, leading to a lack of information available from which to impute the next data point. As the time series moving forward, the amount of previous data increases and helps to improve the imputation accuracy of the Bayesian model. However, when there is a large amount of missing data in the dataset, this approach still does not perform well. Also, its imputation accuracy also depends on the Bayesian model. It is hard to decide whether this model is most suitable for one specific situation or can be generalized. Ultimately, when the situation changes, the Bayesian model may change as well, and thus, it is not generally applicable.

5.5 Exercises

- If you are given 200 vehicle images, would you use CNN? If so, which architecture would you try and why?

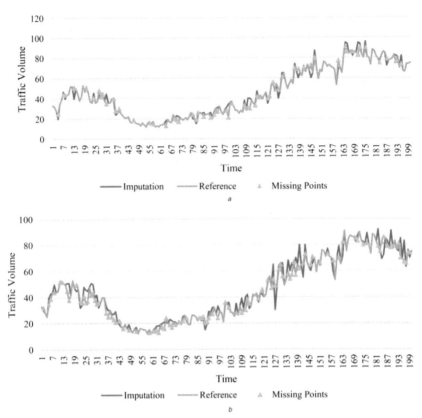

FIGURE 5.18 Imputation results of DSAE approach: (a) 10% Missing ratio, (b) 40% Missing ratio.

- How about if you are given 20,000 images, which CNN models would you try and why?
- What is stride? What is the effect of high stride on the feature map?
- Given input sequence [1, 3, 6, 4, 8, 9, 5] and a filter [1, 0, 1], what is the output of the convolution operation with stride 2 with zero padding?
- What is the size of the output feature map given 7×7 input and two filters of 3×3 with stride 1 and no padding?
- What is the output of 1D max pooling over input [1, 2, 3, 4, 5, 6] with filter size 2 and stride 2?
- What are the different types of pooling? Explain their characteristics.
- List down the hyperparameters of a pooling layer.
- Explain the role of the convolutional layer in CNN.
- Explain the role of the flattening layer in CNN.
- Explain the role of the fully connected (FC) layer in CNN.

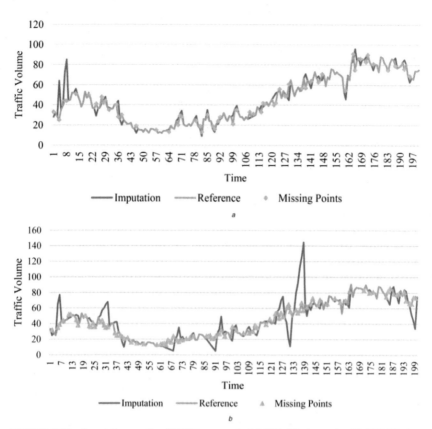

FIGURE 5.19   Imputation results of BPCA approach: (a) 10% Missing ratio, (b) 40% Missing ratio.

- What are the problems associated with the Convolution operation and how can one resolve them?
- Explain the significance of "parameter sharing" and "sparsity of connections" in CNN.
- Let us consider a CNN having three different convolutional layers in its architecture as – Layer-1: Filter Size – 3 X 3, Number of Filters – 10, Stride – 1, Padding – 0 Layer-2: Filter Size – 5 X 5, Number of Filters – 20, Stride – 2, Padding – 0 Layer-3: Filter Size – 5 X5, Number of Filters – 40, Stride – 2, Padding – 0. If we give as input a 3-D image to the network of dimension 39 X 39, then determine the dimension of the vector after passing through a fully connected layer in the architecture.

Chapter 6

Recurrent neural networks

In the previous chapter, we explored the convolutional neural network and a variety of CNN-based model architectures. CNNs mostly focuses on processing image data, and video data has been widely used in transportation applications including traffic object detection, recognition, and classification. However, one of the key properties of transportation-related data, i.e., temporal information, cannot be easily encoded from time series data by feed-forward (fully connected) neural networks and CNNs to resolve the temporal dependency in various transportation tasks, such as traffic flow prediction. Characterizing the temporal dependency in time series is a long-standing but still popular topic since time series data with randomness and periodicity plays an important role in a lot of analytical and forecasting tasks, like weather forecasting and travel planning. Many classical methods and algorithms have been proposed to process time series, including Markov models, spectral analysis methods, autoregressive models, and its popular variant Auto-Regressive Integrated Moving Average (ARIMA). Although these classical models are relatively simple, to some extent, they have similar model structures to the recently developed neural network structure that we will introduce next.

In this chapter, we will focus on another type of neural network structure, recurrent neural network (RNN), which is suitable for dealing with sequential data. RNN and its variants have been widely used in the sequential data modeling, such as natural language processing (NLP) and other advanced neural network architectures. We begin by introducing the basic structure of RNN and two of its variants, i.e., the well-known long short-term memory network (LSTM) and gated recurrent unit network (GRU). One important architecture for capturing the forward and backward temporal dependencies in time series, i.e., the bidirectional structure, is introduced taking the bidirectional LSTM as an example. To learn a representation of time series data and better characterize the dependencies between input and output sequences, the widely used sequence-to-sequence structure, also called encoder–decoder, and a critical attention mechanism are also detailed in this chapter. A more advanced architecture adopting the attention mechanism called transformer, which has nearly reformed the sequential modeling in the deep learning field, is introduced to shed more light on the importance of properly designing neural network models. Following the methodology introduction, two categories of RNN-related transportation tasks are presented to give the reader a sense of how to apply RNNs and related models to the real transportation applications.

Machine Learning for Transportation Research and Applications
https://doi.org/10.1016/B978-0-32-396126-4.00011-4

6.1 RNN fundamentals

RNN is a class of powerful deep neural network that uses its internal memory with loops to deal with sequence data. Unlike CNNs, RNNs can scale to much longer sequences than would be practical for networks without sequence-based specialization. Most RNNs can also process sequences of variable length. The architecture of RNNs is illustrated in Fig. 6.1.

To process the sequence input, RNN takes advantage of one of the early ideas found in machine learning and statistical models of the 1980s: sharing parameters across different parts of a model (Goodfellow et al., 2016). Parameter sharing makes it possible to adapt the model to different forms (different lengths) and generalize across them. If we had separate parameters for each value of the time index in the sequential input, we could not generalize to sequence lengths. The shared parameters are hosted in the RNN cell, as shown in Fig. 6.1, to compute the hidden states h_t. Unlike the introduced fully connected layers and CNNs, RNNs produce an output at each time step and have recurrent connections between the hidden units.

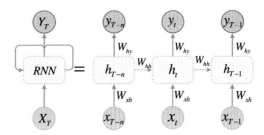

FIGURE 6.1   Standard RNN architecture and an unfolded structure with T time steps.

The RNN cell receives the input vector, X_T, and generates the output, which can be a vector Y_T or the last element of y_T. Here, besides being a value, the component x_t in X_T can also be a vector $x_t \in \mathbb{R}^n$ with n elements. In this case, X_T will be a 2-D matrix $X_T \in \mathbb{R}^{T \times n}$.

The unfolded structure of RNNs, shown on the right in Fig. 6.1, presents the calculation process that, at each time iteration t, the hidden layer maintains a hidden state, h_t, and updates it based on the layer input, x_t, and the previous hidden state, h_{t-1}, using the following equation:

$$h_t = \sigma_h(W_{xh}x_t + W_{hh}h_{t-1} + b_h), \tag{6.1}$$

where W_{xh} is the weight matrix from the input layer to the hidden layer, W_{hh} is the weight matrix between two consecutive hidden states (h_{t-1} and h_t), b_h is the bias vector of the hidden layer, and σ_h is the activation function to generate the hidden state. The network output can be characterized as:

$$y_t = \sigma_y(W_{hy}h_t + b_y), \tag{6.2}$$

where W_{hy} is the weight matrix from the hidden layer to the output layer, b_y is the bias vector of the output layer, and σ_y is the activation function of the output layer. The parameters of the RNN is trained and updated iteratively via the back-propagation through time (BPTT) method. In each time step t, the hidden layer will generate a value, y_t. In some tasks, like single-step traffic prediction, only the last output, y_T, is the desired result.

Although RNNs exhibit the superior capability of modeling nonlinear time series problems (Ma et al., 2015), regular RNNs suffer from the vanishing or blowing up of the gradient during the BPTT process. The reason is that RNN is trained by BPTT, and therefore unfolded into a structure like a feed-forward network with multiple layers. When the gradient is passed back through many time steps, it tends to grow or vanish. Thus, RNNs are incapable of learning from long time lags (Gers et al., 1999), or saying long-term dependencies (Bengio et al., 1994). This section introduces the basic structure of the RNN and describes the shortcomings of RNNs briefly. If you are interested in knowing more about the basic theory underlying RNN and its variants, please refer to Sherstinsky (2020).

6.2 RNN variants and related architectures

6.2.1 Long short-term memory (LSTM) and gated recurrent units (GRU)

To overcome the gradient problem, gate units with nonlinear activation functions and separated cell states are added to the RNN cells to prevent the gradient from vanishing or exploding. A representative RNN variant with gate units is the Long Short-Term Memory (LSTM). It has been shown that LSTMs work well on sequence-based tasks with long-term dependencies (Duan et al., 2016; Chen et al., 2016). Although there are a variety of typical LSTM variants proposed in recent years, a comprehensive analysis of LSTM variants shows that few RNN variants can improve upon the standard LSTM architecture significantly (Greff et al., 2017). Thus, we mainly introduce LSTM as the most representative RNN variant in this subsection.

The structure of LSTM cell, a variant of the RNN hidden unit, is shown in Fig. 6.2. At time step $t \in \{1, 2, ..., T\}$, the LSTM layer maintains a hidden memory cell $\tilde{C}_t$ and three gate units, which are input gate i_t, forget gate f_t, and output gate o_t. The LSTM cell takes the current variable vector x_t, the preceding output h_{t-1}, and preceding cell state C_{t-1} as inputs. With the memory cell and the gate units, LSTM can learn long-term dependencies to allow useful information to pass along the LSTM network. The gate structure, especially the forget gate, helps LSTM to be an effective and scalable model for several learning problems related to sequential data (Greff et al., 2017). The input gate, the forget gate, the output gate, and the memory cell are represented by colored boxes in the LSTM

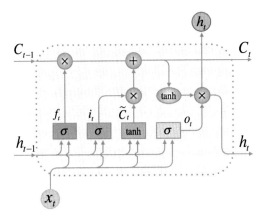

FIGURE 6.2 Structure of the LSTM cell.

cell in Fig. 6.2. Their calculation can be described as the following equations:

$$f_t = \sigma_g(W_f \cdot x_t + U_f \cdot h_{t-1} + b_f) \tag{6.3}$$
$$i_t = \sigma_g(W_i \cdot x_t + U_i \cdot h_{t-1} + b_i) \tag{6.4}$$
$$o_t = \sigma_g(W_o \cdot x_t + U_o \cdot h_{t-1} + b_o) \tag{6.5}$$
$$\tilde{C}_t = \tanh(W_C \cdot x_t + U_C \cdot h_{t-1} + b_C), \tag{6.6}$$

where $\cdot$ is the matrix multiplication operator. W_f, W_i, W_o, and W_C are the weight matrices mapping the hidden layer input to the three gate units and the memory cell, while the U_f, U_i, U_o, and U_C are the weight matrices connecting the preceding output to the three gates and the memory cell. The b_f, b_i, b_o, and b_C are four bias vectors. The gate activation function $\sigma_g(\cdot)$ is the sigmoid function, and the $\tanh(\cdot)$ is the hyperbolic tangent function. Then, the cell output state C_t and the layer output h_t can be calculated as follows:

$$C_t = f_t \odot C_{t-1} + i_t \odot \tilde{C}_t \tag{6.7}$$
$$h_t = o_t \odot \tanh(C_t), \tag{6.8}$$

where $\odot$ is the element-wise matrix multiplication operator.

The final output of an LSTM layer should be a sequence of the layer output at all time steps, represented by $H_T = [h_1, h_2, ..., h_T]$. If only the last element of the output vector h_T is the desired result, i.e., $\hat{x}_{T+1} = h_T$, in the training process, the model's total loss $\mathcal{L}$ at each iteration can be calculated by

$$\mathcal{L} = \text{Loss}(\hat{x}_{T+1} - x_{T+1}) = \text{Loss}(h_T - x_{T+1}), \tag{6.9}$$

where $\text{Loss}(\cdot)$ is the loss function, which is normally the mean square error function for traffic prediction problems.

6.2.2 Bidirectional RNN

In many applications, we want to produce a prediction of y_t that may depend on the whole input sequence. The bidirectional RNNs were invented to address characterizing bidirectional dependencies, and they have been extremely successful in applications, such as handwriting recognition, speech recognition, and bioinformatics. Bidirectional RNNs are also useful in the transportation domain. For example, in traffic forecasting, since traffic patterns have apparent periodicity, traffic state time series not only have chronologically forward dependencies, but also backward relationships.

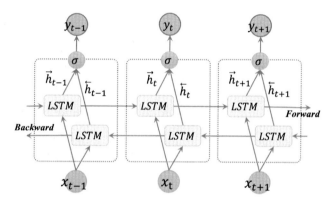

FIGURE 6.3 Bidirectional LSTM.

The idea of the BDLSTM comes from the bidirectional RNN (Schuster and Paliwal, 1997), which processes sequence data in both forward and backward directions with two separate LSTM hidden layers. The structure of an unfolded BDLSTM layer, containing a forward LSTM layer and a backward LSTM layer, is introduced and illustrated in Fig. 6.3. The forward layer output, $\overrightarrow{h}_t$, is iteratively calculated based on positive ordered inputs $[x_1, x_2, ..., x_T]$ and masks $[m_1, m_2, ..., m_T]$. The backward layer output, $\overleftarrow{h}_t$, is iteratively calculated using the reversed ordered inputs and masks from time step T to time step 1. Both forward and backward outputs are calculated based on the LSTM-I model equations, Eqs. (6.3)–(6.8). The BDLSTM layer generates output element y_t at each step t based on $\overrightarrow{h}_t$ and $\overleftarrow{h}_t$ by using the following equation:

$$y_t = \oplus(\overrightarrow{h}_t, \overleftarrow{h}_t), \tag{6.10}$$

where $\oplus$ is an average function to combine the two output sequences in this study. Note that other functions, such as summation, multiply, or concatenate functions, can be used instead. Similar to the LSTM layer, the final output of a BDLSTM layer can be represented by a vector $Y = [y_1, y_2, ..., y_T]$.

6.2.3 Sequence to sequence

The sequence-to-sequence (often abbreviated to seq2seq) model is a special class of RNNs that were typically used in complex language problems like machine translation, question answering, etc. Many seq2seq text understanding tasks rely on the RNN-based an encoder–decoder framework. In such a framework, the encoder first maps an input $\mathbf{x} = (x_1, x_2, \ldots, x_T)$ into a representation vector as the encoder output, and then the decoder RNN takes encoder output as its input and generates the output with multiple steps $\mathbf{y} = (y_1, y_2, \ldots, y_T)$.

Both encoder and the decoder in a seq2seq model are normally LSTM models (or sometimes GRU models). Take LSTM-based seq2seq model as an example: The encoder reads the input sequence and encodes it into the internal state vectors or context vector (in the case of LSTM, these are called the hidden state and cell state vectors). The outputs of the encoder are discarded, and only the internal states are preserved. These internal states, i.e., a context vector c as shown in Fig. 6.4, aims to encapsulate the information for all input elements in order to help the decoder make accurate predictions. There are several ways to incorporate the context vector into the decoder. The simplest way is to take the context vector as the starting input element of the decoder. Another ways is to adopt the attention mechanism, which will be introduced in the next subsection. The LSTM decoder's initial states are initialized as the final states of the encoder LSTM, i.e., the context vector of the encoder's final cell is the input to the first cell of the decoder network. Using these initial states, the decoder starts generating the output sequence, and these outputs are also taken into consideration for future outputs.

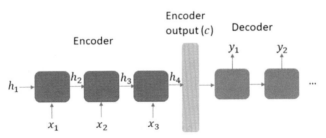

FIGURE 6.4 Sequence-to-sequence (Seq2Seq) architecture.

6.3 RNN as a building block for transportation applications

6.3.1 RNN for road traffic prediction

Problem description

Traffic prediction algorithms can fulfill the prediction of many types of traffic state variables, such as road flow, speed, and travel time. In this section, we take the traffic speed prediction as an example to show how to predict traffic speed using RNNs. Traffic speed prediction at one location normally uses a sequence of speed values with n historical time steps as the input data, which can be represented by the vector:

$$X_T = \left[x_{T-n}, x_{T-(n-1)}, \ldots, x_{T-2}, \ x_{T-1} \right]. \tag{6.11}$$

But the traffic speed at one location may be influenced by the speeds of nearby locations or even locations faraway, especially when a traffic jam propagates through the traffic network. To take these network-wide influences into account, the proposed and compared models in this study take the network-wide traffic speed data as the input. Supposing the traffic network consists of P locations and we need to predict the traffic speeds at time T using n historical time frames (steps), the input can be characterized as a speed data matrix,

$$X_T^P = \begin{bmatrix} x^1 \\ x^2 \\ \vdots \\ x^P \end{bmatrix} = \begin{bmatrix} x_{T-n}^1 & x_{T-n+1}^1 & \cdots & x_{T-2}^1 & x_{T-1}^1 \\ x_{T-n}^2 & x_{T-n+1}^2 & \ddots & x_{T-2}^2 & x_{T-1}^2 \\ \vdots & \vdots & & \vdots & \vdots \\ x_{T-n}^P & x_{T-n+1}^P & \cdots & x_{T-2}^P & x_{T-1}^P \end{bmatrix}, \tag{6.12}$$

where each element x_t^p represents the speed of the t-th time frame at the p-th location. To reflect the temporal attributes of the speed data and simplify the expressions of the equations in the following subsections, the speed matrix is represented by a vector, $X_T^P = \left[x_{T-n}, x_{T-(n-1)}, \ldots, x_{T-2}, x_{T-1} \right]$, in which each element is a vector of the P locations' speed values.

Network-wide traffic prediction

As shown in Fig. 6.5, the short-term network-wide traffic forecasting problem attempts to predict the future traffic state at time $T + 1$, x_{t+1}, based on a sequence of historical traffic state $[x_{t-n+1}, \ldots, x_{t-1}, x_t]$. Thus, the traffic prediction can be considered to design a function $f(X_t)$ and fit it from historical

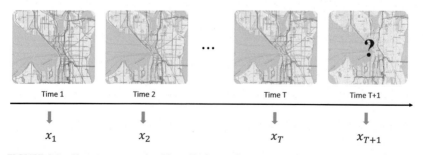

FIGURE 6.5   Short-term network-wide traffic forecasting.

data input X_t to the future traffic states $Y_t = X_{t+1}$. Basically, the traffic forecasting problem can be categorized into two types, single-step forecasting and multistep forecasting, with respect to the number of steps of the future state the model attempts to predict. In this section, we take the single-step prediction as an example. Thus, the traffic state prediction problem aims to learn a function $F(\cdot)$ to map T steps of historical time series to the next subsequent time step of traffic state.

$$F([x_1, x_2, ..., x_t]) = [x_{t+1}]. \tag{6.13}$$

Please note that models with the sequence-to-sequence structure or other advanced architectures will be more suitable for multistep prediction due to its more complicated output with higher dimensions.

Traffic prediction algorithms

The training procedure of RNN-based traffic-prediction algorithm is almost identical to a normal ML training procedure. One potential difference is the training–testing split for algorithm evaluation. The normal model training and evaluation procedure requires shuffling all the data samples and splitting the dataset into three portions, including a training set, a validation set, and a testing set, to ensure all sets follow the identical distribution. However, since the traffic prediction task tends to learn from historical time series and predict future traffic, another data splitting strategy is needed to split the dataset according to the chronological order of the dataset samples where the early portion will be categorized as training set and the latest portion will be the testing set.

In this section, we take the LSTM network as a prediction model to briefly demonstrate procedures for predicting network-wide traffic. Algorithm 1 describes the basic training procedure of the prediction model without showing the details of the LSTM model. To further help you understand the details of the

---

Algorithm 1: Short-term traffic forecasting.

    Data: Training set $\mathcal{D}_{train}$, Validation set $\mathcal{D}_{valid}$, Testing set $\mathcal{D}_{test}$
    Result: Trained Model $LSTM_\omega$
1   Initialize parameters ω_0 in LSTM;
2   **while** *Not Converge* **do**
3   sample (x_t, y_t) from D_{train} ;
4       $\hat{y}_t \leftarrow LSTM(x_t)$;
5   Loss $= error(\hat{y}_t, y_t)$;
6       $\omega_{t+1} \leftarrow \omega_t - r\frac{\partial Loss}{\partial \omega}$;
7   **if** *Loss < Threshold* **then**
8   Return $LSTM_\omega$

---

algorithm, please refer to Exercises 1 and 2 to practice implementing the LSTM model and the network-wide traffic-prediction algorithm.

It also should be noted that the input of a traffic prediction model could be far more than just a combination of the traffic state matrix X_T and the masking matrix M_T. A traffic prediction model can also take many other influential factors into consideration, such as road network structure, temporal information (time of day, day of the week, etc.), weather information, and event information. Thus, the definition of traffic prediction problem explicitly depends on the design of the prediction model.

6.3.2 Traffic prediction with missing values

Problem definition

In reality, traffic sensors, like inductive-loop detectors, may fail due to the breakdown of wire insulation or damage caused by construction activities or electronics unit failure. The sensor failure further causes missing values in collected time series data, as shown by the black boxes in Fig. 6.6. For the RNN-based prediction problem, if the input time series contains missing/null values, the model may fail due to null values that cannot be computed during the training process. If the missing values are set as some predefined values, like zeroes, the means of historical observations, or last observed values, these biased model inputs will result in biased parameter estimation in the training processing (Che et al., 2018). Thus, how to deal with missing values in input sequences needs to be take into consideration in the model design process.

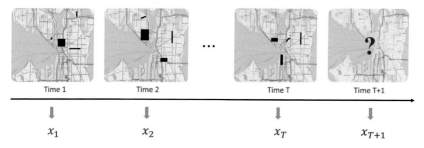

FIGURE 6.6 Short-term network-wide traffic forecasting with missing values in input data.

To formulate this problem, a *masking* vector $m_t \in \{0, 1\}^D$ is adopted to denote whether traffic states are missing at time step t. The masking vector for x_t is defined as

$$m_t^d = \begin{cases} 1, & \text{if } x_t^d \text{ is observed} \\ 0, & \text{otherwise.} \end{cases} \tag{6.14}$$

Thus, the masking matrix for a data sample is $M = \{m_1, m_2, ..., m_T\}^T \in \mathbb{R}^{T \times D}$.

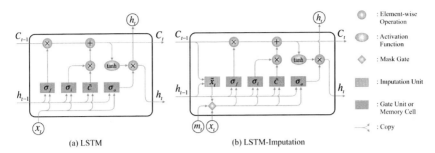

FIGURE 6.7 (a) Structure of LSTM. (b) Structure of LSTM-I.

With the marking definition, the traffic state prediction problem becomes learning a function $F(\cdot)$ to map T time steps of historical time series to the next subsequent time steps of traffic speed information:

$$F([x_1, x_2, ..., x_T]; [m_1, m_2, ..., m_T]) = [x_{T+1}]. \tag{6.15}$$

LSTM-based traffic prediction with missing values

In this subsection, the application of traffic prediction with missing values is employed as an example still based on the LSTM method. To fill missing values and fulfill traffic prediction, Che et al. (2018) proposed a GRU-D method targeting on inferring missing values based on the historical mean and the last observation with a learnable decay rate. In this example, an LSTM-based model, LSTM-I, with an imputation unit for inferring missing values is introduced (Cui et al., 2020). Fig. 6.7 (a) and (b) illustrate the difference in the model structures between the LSTM and LSTM-I.

Specifically, the imputation unit σ_p is fed with the preceding cell state C_{t-1} and the preceding output h_{t-1} to infer the values of the subsequent observation, as shown in Fig. 6.7 (b). The inferred observation $\tilde{x}_t \in \mathbb{R}^D$ is denoted as

$$\tilde{x}_t = \sigma_p(W_I \cdot C_{t-1} + U_I \cdot h_{t-1} + b_I), \tag{6.16}$$

where W_I and U_I are the weights and b_I is the bias in the imputation unit. The values inferred from the imputation values can contribute to the training process. In this way, the LSTM-I can complete the data imputation and prediction tasks at the same time. Thus, it is particularly suitable for online traffic prediction problems, which may commonly encounter missing values issues. Please note that the inferred values may not be the "actual" missing values since the proposed imputation unit is only designed for generating appropriate values to help the calculation process in the LSTM structure work properly and generate accurate predictions.

Then, each missing element of the input vector is updated by the imputed element

$$x_t^d \leftarrow m_t^d x_t^d + \left(1 - m_t^d\right)\tilde{x}_t^d, \tag{6.17}$$

where $\tilde{x}_t^d$ is the d-th element of $\tilde{x}_t$. According to Eq. (6.17), if x_t^d is missing, m_t^d is zero and x_t^d is imputed by $\tilde{x}_t^d$.

Additionally, because each masking vector m_t contains the position information of missing values at time step t, the masking vector is also fed into the model and the LSTM-I structure can be characterized as

$$x_t = m_t \odot x_t + (1 - m_t) \odot \tilde{x}_t. \tag{6.18}$$

In this way, the input x_t of the LSTM-I with all missing values filled can generate prediction output such as the vanilla LSTM model. For more information about the LSTM-I model, please refer to (Cui et al., 2020).

6.4 Exercises

6.4.1 Questions

1. Are there any dimension limitations on the input of RNN models? Why?
2. What are the main reasons to add the gate units in LSTM models?
3. Why is GRU faster as compared to LSTM?
4. Explain the difference between the attention mechanism in Seq2Seq models and transformer.
5. How is the transformer architecture better than RNN?

6.4.2 Project: predicting network-wide traffic using LSTM

In this exercise, you need to complete the provided code to train a LSTM model to predict the network-wide traffic state in Seattle area. Specifically, you need to implement the algorithm, fine-tune the model parameters, and evaluate the model performance.

Problem definition

As introduced in Sect. 6.3.1, the network-wide traffic prediction problem can be characterized as learning a function $F(\cdot)$ to mapping the input historical spatiotemporal traffic data and future traffic state.

Dataset preparation

In this study, a loop detector dataset,[1] collected by inductive loop detectors deployed on roadway surface, is utilized to carry out experiments to test the proposed model. Multiple loop detectors are connected to a detector station deployed around every half a mile. The collected data from each station are grouped and aggregated as station-based traffic-state data according to directions. This aggregated and quality controlled dataset contains traffic speed, volume, and occupancy information. In the experiments, the loop detector data

---

[1] https://github.com/zhiyongc/Seattle-Loop-Data.

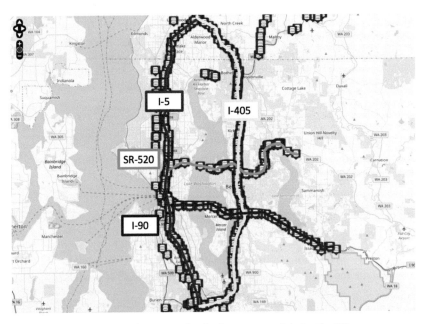

FIGURE 6.8 Loop detector dataset covering the freeway network in Seattle, WA.

cover four connected freeways, which are I-5, I-405, I-90 and SR-520 in the Seattle area, and are extracted from the Digital Roadway Interactive Visualization and Evaluation Network (DRIVE Net) system. The traffic sensor stations, represented by small blue icons, are shown in Fig. 6.8. This dataset contains traffic state data from 323 sensor stations in 2015, and the time-step interval of this dataset is 5 min.

The data will be split into three datasets, i.e., the training set, validation set, and testing set, normally with a splitting ratio of 7:2:1.

Implement and fine-tune model

Exercise 1 shows the details of implementing an LSTM algorithm. Since RNN and its variants are widely used, general deep learning packages, like PyTorch and TensorFlow, provide algorithm packages to enable the user to call for the RNN models using simple API. Thus, in this exercise, you will be required to use a LSTM model provided by PyTorch and implement the all the training, validation, and testing procedures. Similar to Exercise 1, the basic coding structure is provided in this exercise. You will need to fill your own code and fine-tune several hyperparameters, including but not limited to the dimension of the LSTM hidden state, batch size, and learning rate. The exercise coding script is provided in the book's coding repository as well at https://github.com/Deep-Learning-for-Transportation/DL_Transportation.

Model evaluation

After implementing the training and validation procedures of the LSTM-based traffic performance, you need to evaluate the prediction performance on the testing dataset. There are several common prediction performance-measurement metrics, including mean squared error (MSE) and mean absolute percentage error (MAPE). To complete this exercise, please use the well-trained model to evaluate the performance to ensure prediction MAPE $\leq$ 5%. Please note that those deep learning packages all provide the RNN model and many other RNN variants, and you are welcome to select and test other RNN-based models.

Chapter 7

Reinforcement learning

Reinforcement learning (RL) is the area of machine learning that deals with sequential decision-making. In this chapter, we describe how the RL problem can be formalized as an agent that has to make decisions in an environment to optimize a given notion of cumulative rewards. This chapter also introduces the various approaches to learning sequential decision-making tasks and how deep RL can solve a series of decision-making problems in the transportation field.

A key aspect of RL is that an agent learns good behavior. This means that it modifies or acquires new behaviors and skills incrementally. Another important aspect of RL is that it uses trial-and-error experience (as opposed to, e.g., dynamic programming, that assumes full knowledge of the environment a priori). Thus, the RL agent does not require complete knowledge or control of the environment; it only needs to be able to interact with the environment and collect information. Thus, when controlling transportation-related agents, such as vehicles and traffic signal lights, RL-based control strategy can be learned from simulated traffic environments. This chapter shows examples of how the decisions of vehicles and traffic signals are made with the help of the RL algorithms.

7.1 Reinforcement learning setting

The general RL problem is formalized as a discrete-time stochastic-control process where an agent interacts with its environment in the following way: The agent starts, in a given state within its environment $s_0 \in \mathcal{S}$, by gathering an initial observation $\omega_0 \in \Omega$. At each time step t, the agent has to take an action $a_t \in \mathcal{A}$. As illustrated in Fig. 7.1, it follows three consequences: (i) The agent obtains a reward $r_t \in \mathcal{R}$, (ii) the state transitions to $s_{t+1} \in \mathcal{S}$, and (iii) the agent obtains an observation $\omega_{t+1} \in \Omega$. Here, we review the main elements of RL before delving into deep RL in the following sections.

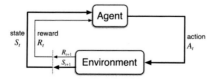

FIGURE 7.1   Agent-environment interaction in RL.

Machine Learning for Transportation Research and Applications
https://doi.org/10.1016/B978-0-32-396126-4.00012-6

7.1.1 Markov property

The Markov property means that the future of the process only depends on the current observation, and the agent has no interest in looking at the full history. A Markov Decision Process (MDP) is a discrete-time stochastic control process that is defined in the following paragraph.

An MDP is a 5-tuple $(\mathcal{S}, \mathcal{A}, P, R, \gamma)$ where:

- $\mathcal{S}$ is the state space and $s_t \in \mathcal{S}$ represents the state of the agent at time t;
- $\mathcal{A}$ is the action space, in which $a_t \in \mathcal{A}$ is the action of the agent following a policy $a_t \sim \pi(s_t, \cdot)$;
- $P : \mathcal{S} \times \mathcal{A} \times \mathcal{S} \to [0, 1]$ is the transition probability function (set of conditional transition probabilities between states), which is normally represented by $P(s_{t+1} \mid s_t, a_t) = T(s_t, a_t, s_{t+1})$;
- $R : \mathcal{S} \times \mathcal{A} \times \mathcal{S} \to \mathcal{R}$ is the reward function, where $\mathcal{R}$ is a continuous set of possible rewards in a range $R_{\max} \in \mathbb{R}^+$ (e.g., $[0, R_{\max}]$);
- $\gamma \in [0, 1)$ is the discount factor.

If the system is fully observable in an MDP, the observation is the same as the state of the environment: $\omega_t = s_t$. At each time step t, the probability of moving to s_{t+1} is given by the state transition function $P(s_{t+1}|s_t, a_t)$, and the reward is given by a bounded reward function $R(s_t, a_t, s_{t+1}) \in \mathcal{R}$.

7.1.2 Goal of reinforcement learning

The goal of an RL agent is to find a policy $\pi(s, a) \in \Pi$, so as to optimize an expected return $V^\pi(s) : \mathcal{S} \to \mathbb{R}$ (also called V-value function) such that

$$V^\pi(s) = \mathbb{E}\left[\sum_{k=0}^{\infty} \gamma^k r_{t+k} \mid s_t = s, \pi\right], \tag{7.1}$$

where the reward can be described as $r_t = \mathbb{E}_{a \sim \pi(s_t, \cdot)} R(s_t, a, s_{t+1})$. From the definition of the expected return, the optimal expected return can be defined as finding a policy $\pi \in \Pi$ to achieve the maximum expected return:

$$V^*(s) = \max_{\pi \in \Pi} V^\pi(s). \tag{7.2}$$

In addition, RL also introduces a few other functions of interest to achieve its goal. The state-action (Q-value) function $Q^\pi(s, a) : \mathcal{S} \times \mathcal{A} \to \mathbb{R}$ is defined as follows:

$$Q^\pi(s, a) = \mathbb{E}\left[\sum_{k=0}^{\infty} \gamma^k r_{t+k} \mid s_t = s, a_t = a, \pi\right]. \tag{7.3}$$

Based on Bellman's equation, Eq. (7.3) can be rewritten recursively as

$$Q^\pi(s, a) = \sum_{s' \in \mathcal{S}} P(s'|a, s) \left(R(s, a, s') + \gamma Q^\pi(s', a = \pi(s'))\right). \tag{7.4}$$

Similar to the value function V, the optimal Q function $Q^*(s, a)$ can also be defined as

$$Q^*(s, a) = \max_{\pi \in \Pi} Q^\pi(s, a). \tag{7.5}$$

The difference between the V-value function and the Q-value function is that the optimal V-value function $V^*(s)$ is the expected discounted reward given a state s when the agent follows the optimal policy π^*, while the optimal Q-value $Q^*(s, a)$ is the expected discounted return given a state s with a given action a when agent also follows the optimal policy π^* thereafter. Particularly, the optimal policy can be obtained directly from the Q-value function $Q^*(s, a)$:

$$\pi^*(s) = \underset{a \in \mathcal{A}}{\operatorname{argmax}} Q^*(s, a). \tag{7.6}$$

Sometimes, RL will use another form of the expected return, which is the **advantage** function defined

$$A^\pi(s, a) = Q^\pi(s, a) - V^\pi(s). \tag{7.7}$$

This quantity of $A^\pi(s, a)$ describes how good the action a is, compared to the expected return when directly following the policy π.

7.1.3 Categories and terms in reinforcement learning

Before presenting the details of RL algorithms, we will introduce a pair of terms that are usually mentioned in RL.

- Episode: Each subsequence of agent–environment interactions between initial and terminal states is called an episode. Taking a simulated environment as an example, an episode is a simulation event that ends in a terminal state. Consider a video game, the time your player is alive is an episode. Episodic tasks in RL mean that the game ends at a terminal stage or after some amount of time. Whenever an episode ends, the game reverts to the initial state (not necessarily the same all the time).
- Horizon: A horizon in reinforcement learning is a future point relative to a time step, beyond which you do not care about reward (so you sum the rewards from time t to $t + H$). The finite horizon is the agent's time to live.

In the following subsections, several categories of RL settings, policies, training strategies, etc., are introduced to help you understand how RL is trying to solve what kind of problems.

Model-free vs. model-based

The goal of RL is to find a policy to maximize the reward that the agent attain. There are several necessary components to learn a policy for an agent: (1) a representation of a value function that provides a prediction of how good each state or each state/action pair is; (2) a direct representation of the policy $\pi(s)$, or $\pi(s, a)$, or (3) a model of the environment (the estimated transition function and the estimated reward function) in conjunction with a planning algorithm. The first two components are related to what is called model-free RL. When the latter component is used, the algorithm is referred to as model-based RL.

- **Model-free** learning is a simple RL process in which a value is associated with actions. The agent does not need to build a model to describe the internal operation mechanism of the surrounding environment.
- **Model-based** learning, as it sounds, tries to understand the agent's environment and creates a model for it based on its interactions with this environment. Model-based RL relies on the formation of internal models of the environment to maximize the reward.

Thus, model-free and model-based are two categories of RL based on whether the agent needs to understand the environment or not. Most deep learning-based RL algorithms are model-free algorithms since deep learning methods are inherently black-box methods. These methods mostly build end-to-end neural networks to generate the actions of the agents without modeling the environment. It is also worth noting that, in most real-world RL problems, the state space is high-dimensional (and possibly continuous), and neural networks have advantages for RL algorithms since they are well suited to deal with high-dimensional inputs (such as times series, frames, etc.).

Stationary vs. nonstationary

In RL, a policy defines how an agent determines actions based on states. The policy is learned based on the interaction between the agents and the environment. Stationary and nonstationary refer to whether the environment keeps changing or not. A stationary (time) series is one whose statistical properties, such as the mean, variance, and auto-correlation, are all constant over time. On the other hand, a nonstationary series is one whose statistical properties change over time.

Thus, policies can be categorized under the criterion of being either **stationary** or **nonstationary as follows**.

- A nonstationary policy depends on the time-step and is useful for the finite horizon context where the cumulative rewards that the agent seeks to optimize are limited to a finite number of future time steps. Here, you can regard the finite horizon as the agent's time to live.
- A stationary policy is a policy that does not change. It generally means that the policy is not being updated by a learning algorithm. If the infinite horizons are considered, then the policies are considered stationary.

Deterministic policy vs. stochastic policy

In terms of the manner for an agent to determine the action based on its state, policies can be categorized under another criterion of being either **deterministic** or **stochastic**:

- In the deterministic case, the policy is described by $\pi(s) : S \rightarrow A$.
- In the stochastic case, the policy is described by $\pi(s, a) : S \times A \rightarrow [0, 1]$, where $\pi(s, a)$ denotes the probability that action a may be chosen in state s.

The formalism can be directly extended to the finite horizon context. In that case, the policy and the cumulative expected returns should be time-dependent.

Offline learning vs. online learning

In terms of the different settings to learn a policy from data, the RL settings can be categorized as **offline learning** and **online learning**. Learning a sequential decision-making task, like MDP, appears in two cases: (i) in the offline learning case where only limited data on a given environment is available and (ii) in an online learning case where, in parallel to learning, the agent gradually gathers experience in the environment.

- In an offline setting, the experience is acquired a priori, and then it is used as a batch for learning; hence the offline setting is also called batch RL. Strictly offline learning algorithms need to be rerun from scratch in order to learn from the changed data.
- In contrast, in the online setting, data becomes available in sequential order and is used to progressively update the behavior of the agent. Strictly online algorithms improve incrementally as each piece of new data arrives, then discard that data and do not use it again. It is not a requirement, but it is commonly desirable for an online algorithm to forget older examples over time, so that it can adapt to nonstationary populations.

The main difference between online and offline is that, in an online setting, the agent can influence how it gathers experience so that it can be more useful for learning. However, it is also an additional challenge for the agent since it has to deal with the exploration/exploitation dilemma while learning.

Exploration vs. exploitation

The more episodes are collected the better because this improves the estimates of the function. However, there's a problem. If the algorithm for policy improvement always updates the policy greedily, meaning it takes only actions leading to immediate reward, actions and states not on the greedy path will not be sampled sufficiently, and potentially better rewards would stay hidden from the learning process.

Essentially, we're forced to make a choice between making the best decision given the current information or starting to explore and find more information. This is also known as the exploration vs. exploitation dilemma.

In fact, we're looking for something like a middle ground between those. Full-on exploration would mean that we would need a lot of time to collect the needed information, and full-on exploitation would cause the agent to get stuck in a local reward maximum. There are two approaches to ensure all actions are sampled sufficiently called on-policy and off-policy methods.

Off-policy learning vs. on-policy learning

In terms of how to learn experience from the environment, RL learning strategies can be categorized as **off-policy** and **on-policy**. In RL, there could be two terms that are related to off-policy and on-policy learning:

- Target Policy $\pi(s, a)$: It is the policy that an agent is trying to learn, i.e., the agent is learning the value function for this policy.
- Behavior Policy $\beta(s, a)$: It is the policy that is being used by an agent for action select, i.e., the agent follows this policy to interact with the environment.

The difference between off-policy and on-policy can be described as follows:

- On-policy learning algorithms are the algorithms that evaluate and improve the same policy that is being used to select actions. That means we will try to evaluate and improve the same policy that the agent is already using for action selection. In short, *target policy == behavior policy*. On-policy based methods usually introduce a bias when used with a replay buffer because the trajectories are usually not obtained solely under the current policy π. Some examples of on-policy algorithms are policy iteration, value iteration, Monte Carlo for on-policy, SARSA, etc.
- Off-policy learning algorithms evaluate and improve a policy that is different from the policy that is being used for action selection. In off-policy based methods, learning is straightforward when using trajectories that are not necessarily obtained under the current policy, but from a different behavior policy $\beta(s, a)$. In short, *target policy $\neq$ behavior policy*. In those cases, experience replay allows reusing samples from a different behavior policy. Some examples of off-policy learning algorithms are Q learning, expected SARSA (can act in both ways), etc.

More specifically, this difference in learning strategy makes off-policy methods sample efficient as they are able to make use of any experience. The benefits of off-policy methods are as follows:

- Continuous exploration: As an agent is learning an other policy, it can then be used for continuing exploration, while learning optimal policy, whereas on-policy learns suboptimal policy.
- Learning from demonstration: an agent can learn from the demonstration.
- Parallel learning: This speeds up the convergence, i.e., learning can be fast.

7.2 Value-based methods

Value-based algorithms aim to build a value function that can subsequently define a policy for the agent. The Q-learning related algorithms (Watkins and Dayan, 1992) are the most popular value-based algorithms. In this section, we will also specifically discuss the main elements of the deep Q-network (DQN) algorithm (Mnih et al., 2015), which uses neural networks as function approximators and has achieved super-human-level control performance when playing video games and controlling other simulated agents. We will also briefly introduce improvements of the DQN algorithm.

7.2.1 Q-learning

The basic version of Q-learning maintains a lookup table of values $Q(s, a)$ with one entry for every state–action pair. In order to learn the optimal Q-value function, the Q-learning algorithm makes use of the Bellman equation for the Q-value function (Dreyfus and Bellman, 1962) whose unique solution is $Q^*(s, a)$:

$$Q^*(s, a) = \left(\mathcal{B}Q^*\right)(s, a), \tag{7.8}$$

where $\mathcal{B}$ is the Bellman operator mapping any function $K : \mathcal{S} \times \mathcal{A} \to \mathbb{R}$ into another function $\mathcal{S} \times \mathcal{A} \to \mathbb{R}$ and is defined as follows:

$$(\mathcal{B}K)(s, a) = \sum_{s' \in \mathcal{S}} T\left(s, a, s'\right) \left(R\left(s, a, s'\right) + \gamma \max_{a' \in \mathcal{A}} K\left(s', a'\right)\right). \tag{7.9}$$

Here, the K function can be any function. Taking the Q function as an example, this function is similar to Eq. (7.4) with the same structure. In practice, one general proof of convergence to the optimal value function is available under the conditions that: (1) the state–action pairs are represented discretely, and (2) all actions are repeatedly sampled in all states (which ensures sufficient exploration, hence not requiring access to the transition model). This simple setting is often inapplicable due to the high-dimensional (possibly continuous) state–action space.

Q-Learning algorithms learn from experiences a given dataset D in the form of tuples $< s, a, r, s' >$ where the state at the next time-step s' is drawn from $T(s, a, \cdot)$ and the reward r is given by $R\left(s, a, s'\right)$. In fitted Q-learning, the algorithm starts with some random initialization of the Q-values $Q\left(s, a; \theta_0\right)$, where θ_0 refers to the initial parameters. Then, an approximation of the Q-values at the kth iteration $Q\left(s, a; \theta_k\right)$ is updated towards the target value

$$Y_k^Q = r + \gamma \max_{a' \in \mathcal{A}} Q\left(s', a'; \theta_k\right), \tag{7.10}$$

where θ_k refers to some parameters that define the Q-values at the kth iteration.

In neural fitted Q-learning (NFQ), the Q function is a fitted neural network where its state can be provided as an input to the Q-network and a different output is given for each of the possible actions. This provides an efficient structure that has the advantage of obtaining the computation of $\max_{a' \in A} Q\left(s', a'; \theta_k\right)$ in a single forward pass in the neural network for a given s'. The Q values are parameterized $Q\left(s, a; \theta_k\right)$, where the parameters θ_k can be updated by stochastic gradient descent by minimizing the square loss:

$$L_{DQN} = \left(Q\left(s, a; \theta_k\right) - Y_k^Q\right)^2.$$
(7.11)

Thus, the Q-learning update amounts in updating the parameters

$$\theta_{k+1} = \theta_k + \alpha \left(Y_k^Q - Q\left(s, a; \theta_k\right)\right) \nabla_{\theta_k} Q\left(s, a; \theta_k\right),$$
(7.12)

where α is the learning rate. Calculating the square loss in Eq. (7.11) ensures that $Q\left(s, a; \theta_k\right)$ should tend without bias to the expected value of the random variable Y_k^Q. Hence, it ensures that $Q\left(s, a; \theta_k\right)$ should tend to $Q^*(s, a)$ after many iterations. Neural networks are well-suited for this kind of end-to-end function-fitting task with sufficient experience/samples gathered in the dataset D.

Early Q-learning methods only use linear functions to fit the Q functions, which will lead to large errors at multiple places in high-dimensional state–action spaces. Although kernel and other methods can add nonlinear properties to the function, this approach is not enough to guarantee convergence. Neural networks have certain generalization and extrapolation abilities, but specific care still has to be taken to ensure proper learning because of the instabilities and the risk of overestimation.

7.2.2 Deep Q-networks

Deep Q-network (DQN) introduced by (Mnih et al., 2015) is probably the most famous reinforcement learning method because it achieves superior performance in a variety of games. It uses two heuristics to improve the algorithm's stability:

- The target Q-network described by Eq. (7.10) is replaced by $Q\left(s', a'; \theta_k^-\right)$, where its parameters θ_k^- are updated only every $C \in \mathbb{N}$ iterations with the following assignment: $\theta_k^- = \theta_k$. This prevents the instabilities to propagate quickly, and it reduces the risk of divergence as the target values Y_k^Q are kept fixed for C iterations. The idea of target networks can be seen as an instantiation of fitted Q-learning, where each period between target network updates corresponds to a single-fitted Q-iteration.
- In an online setting, the replay memory keeps all information for the last $N_{\text{replay}} \in \mathbb{N}$ time steps, where the experience is collected by following an ϵ-greedy policy. The updates are then made on a set of tuples $\langle s, a, r, s'\rangle$ (called

mini-batch) selected randomly within the replay memory. This technique allows for updates that cover a wide range of the state–action space. In addition, one mini-batch update has less variance compared to a single tuple update. Consequently, it provides the possibility to make a larger update of the parameters, while having an efficient parallelization of the algorithm.

In addition to the target Q-network and the replay memory, DQN uses other important heuristics. To keep the target values in a reasonable scale and to ensure proper learning in practice, rewards are clipped between -1 and $+1$. Clipping the rewards limits the scale of the error derivatives and makes it easier to use the same learning rate across multiple games (however, it introduces a bias). In games where the player has multiple lives, one trick is also to associate a terminal state to the loss of a life, such that the agent avoids these terminal states (in a terminal state the discount factor is set to 0).

In DQN, many deep learning specific techniques are also used. In particular, a preprocessing step of the inputs is used to reduce the input dimensionality, to normalize inputs (it scales pixels value into $[-1, 1]$), and to deal with some specificities of the task. In addition, convolutional layers are used for the first layers of the neural network function approximator, and the optimization is performed using a variant of stochastic gradient descent called RMSprop (Tieleman et al., 2012).

7.3 Policy gradient methods for deep RL

This section focuses on a particular family of reinforcement learning algorithms that use policy gradient methods. These methods optimize a performance objective (typically the expected cumulative reward) by finding a good policy (e.g., a neural network parameterized policy) thanks to variants of stochastic gradient ascent with respect to the policy parameters. Note that policy gradient methods belong to a broader class of policy-based methods that includes, among others, evolution strategies. These methods use a learning signal derived from sampling instantiations of policy parameters and the set of policies is developed towards policies that achieve better returns.

In this chapter, we introduce the stochastic and deterministic gradient theorems that provide gradients on the policy parameters in order to optimize the performance objective. Then, we present different RL algorithms that make use of these theorems.

7.3.1 Stochastic policy gradient

The expected return of a stochastic policy π starting from a given state s_0 from Eq. (7.1) can be written as:

$$V^\pi(s_0) = \int_{\mathcal{S}} \rho^\pi(s) \int_{\mathcal{A}} \pi(s, a) R'(s, a) da \, ds, \qquad (7.13)$$

where $R'(s, a) = \int_{s' \in \mathcal{S}} T\left(s, a, s'\right) R\left(s, a, s'\right)$ and $\rho^\pi(s)$ is the discounted state distribution defined as

$$\rho^\pi(s) = \sum_{t=0}^{\infty} \gamma^t \Pr\{s_t = s \mid s_0, \pi\}. \tag{7.14}$$

For a differentiable policy π_w, the fundamental result underlying these algorithms is the policy gradient theorem (Sutton et al., 2000):

$$\nabla_w V^{\pi_w}(s_0) = \int_{\mathcal{S}} \rho^{\pi_w}(s) \int_{\mathcal{A}} \nabla_w \pi_w(s, a) Q^{\pi_w}(s, a) da \, ds. \tag{7.15}$$

This result enables us to adapt the policy parameters $w : \Delta w \propto \nabla_w V^{\pi_w}(s_0)$ from experience. This result is particularly interesting since the policy gradient does not depend on the gradient of the state distribution (even though one might have expected it to). The simplest way to derive the policy gradient estimator (i.e., estimating $\nabla_w V^{\pi_w}(s_0)$ from experience) is to use a score function gradient estimator, commonly known as the REINFORCE algorithm (Williams, 1992). The likelihood ratio trick can be exploited as follows to derive a general method of estimating gradients from expectations:

$$\begin{aligned} \nabla_w \pi_w(s, a) &= \pi_w(s, a) \frac{\nabla_w \pi_w(s, a)}{\pi_w(s, a)} \\ &= \pi_w(s, a) \nabla_w \log\left(\pi_w(s, a)\right). \end{aligned} \tag{7.16}$$

Considering Eq. (7.16), it follows that

$$\nabla_w V^{\pi_w}(s_0) = \mathbb{E}_{s \sim \rho^{\pi_w}, a \sim \pi_w}\left[\nabla_w \left(\log \pi_w(s, a)\right) Q^{\pi_w}(s, a)\right]. \tag{7.17}$$

Note that, in practice, most policy gradient methods effectively use undiscounted state distributions, without hurting their performance.

So far, we have shown that policy gradient methods should include a policy evaluation followed by a policy improvement. On the one hand, the policy evaluation estimates Q^{π_w}. On the other hand, the policy improvement takes a gradient step to optimize the policy $\pi_w(s, a)$ with respect to the value function estimation. Intuitively, the policy improvement step increases the probability of the actions proportionally to their expected return.

The question that remains is how the agent can perform the policy evaluation step, i.e., how to obtain an estimate of $Q^{\pi_w}(s, a)$. The simplest approach to estimating gradients is to replace the Q function estimator with a cumulative return from entire trajectories. In the Monte Carlo policy gradient, we estimate the $Q^{\pi_w}(s, a)$ from rollouts on the environment while following policy π_w. The Monte Carlo estimator is an unbiased well-behaved estimate when used in conjunction with the back-propagation of a neural network policy because it estimates returns until the end of the trajectories (without instabilities induced by bootstrapping). However, the main drawback is that the estimate requires on-policy rollouts and can exhibit high variance. Several rollouts are typically needed to obtain a good estimate of the return. A more efficient approach is to instead use an estimate of the return given by a value-based approach, as in actor–critic methods discussed in § 7.4.1.

We make two additional remarks. 1) To prevent the policy from becoming deterministic, it is common to add an entropy regularizer to the gradient. With this regularizer, the learned policy can remain stochastic. This ensures that the policy keeps exploring. 2) Instead of using the value function Q^{π_w} in Eq. (7.17), an advantage value function A^{π_w} can also be used. While $Q^{\pi_w}(s, a)$ summarizes the performance of each action for a given state under policy π_w, the advantage function $A^{\pi_w}(s, a)$ provides a measure of comparison for each action to the expected return at the state s, given by $V^{\pi_w}(s)$. Using $A^{\pi_w}(s, a) = Q^{\pi_w}(s, a) - V^{\pi_w}(s)$ usually produces lower magnitudes than $Q^{\pi_w}(s, a)$. This helps reduce the variance of the gradient estimator $\nabla_w V^{\pi_w}(s_0)$ in the policy improvement step, while not modifying the expectation [1]. In other words, the value function $V^{\pi_w}(s)$ can be seen as a baseline or control variate for the gradient estimator. When updating the neural network that fits the policy, using such a baseline allows for improved numerical efficiency, i.e., reaching a given performance with fewer updates, because the learning rate can be higher.

7.3.2 Deterministic policy gradient

The policy gradient methods may be extended to deterministic policies. The Deep Deterministic Policy Gradient (DDPG) algorithm introduces the direct representation of a policy in such a way that it can extend the DQN algorithm to overcome the restriction of discrete actions.

Let us denote by $\pi(s)$ the deterministic policy: $\pi(s) : \mathcal{S} \to \mathcal{A}$. In discrete action spaces, a direct approach is to build the policy iteratively with

$$\pi_{k+1}(s) = \underset{a \in \mathcal{A}}{\operatorname{argmax}} Q^{\pi_k}(s, a), \qquad (7.18)$$

where π_k is the policy at the kth iteration. In continuous action spaces, a greedy policy improvement becomes problematic, requiring a global maximization at every step. Instead, let us denote by $\pi_w(s)$ a differentiable deterministic policy. In that case, a simple and computationally attractive alternative is to move the policy in the direction of the gradient of Q, which leads to the Deep Deterministic Policy Gradient (DDPG) (Lillicrap et al., 2015) algorithm:

$$\nabla_w V^{\pi_w}(s_0) = \mathbb{E}_{s \sim \rho^{\pi_w}} \left[\nabla_w (\pi_w) \nabla_a \left(Q^{\pi_w}(s, a) \right) \big|_{a = \pi_w(s)} \right]. \qquad (7.19)$$

This equation implies relying on $\nabla_a (Q^{\pi_w}(s, a))$ (in addition to $\nabla_w \pi_w$), which usually requires using actor–critic methods (see § 7.4.1).

Indeed, subtracting a baseline that only depends on s to $Q^{\pi_w}(s, a)$ in Eq. (7.15) does not change the gradient estimator because $\forall s, \int_{\mathcal{A}} \nabla_w \pi_w(s, a) da = 0$.

7.4 Combining policy gradient and Q-learning

Policy gradient is an efficient technique for improving a policy in a reinforcement learning setting. As we have seen, this typically requires an estimate of a value function for the current policy, and a sample efficient approach is to use an actor–critic architecture that can work with off-policy data.

These algorithms have the following properties unlike the DQN-based methods:

1. They are able to work with continuous action spaces. This is particularly interesting in applications such as robotics, where forces and torques can assume a continuum of values.
2. They can represent stochastic policies, which is useful for building policies that can explicitly explore. This is also useful in settings where the optimal policy is a stochastic policy (e.g., in a multiagent setting where the Nash equilibrium is a stochastic policy).

However, another approach is to combine policy gradient methods directly with off-policy Q-learning. In some specific settings, depending on the loss function and the entropy regularization used, value-based methods and policy-based methods are equivalent. For instance, when adding an entropy regularization, Eq. (7.17) can be written as

$$\nabla_w V^{\pi_w}(s_0) = \mathbb{E}_{s,a}\left[\nabla_w \left(\log \pi_w(s, a)\right) Q^{\pi_w}(s, a)\right] + \alpha \mathbb{E}_s \nabla_w H^{\pi_w}(s), \quad (7.20)$$

where $H^{\pi}(s) = -\sum_a \pi(s, a) \log \pi(s, a)$. From this, one can observe that an optimum is satisfied by the following policy: $\pi_w(s, a) = \exp(A^{\pi_w}(s, a)/\alpha - H^{\pi_w}(s))$. Therefore, we can use the policy to derive an estimate of the advantage function: $\tilde{A}^{\pi_w}(s, a) = \alpha \left(\log \pi_w(s, a) + H^{\pi}(s)\right)$. We can thus think of all model-free methods as different facets of the same approach.

One remaining limitation is that both value-based and policy-based methods are model-free, and they do not make use of any model of the environment. The next chapter describes algorithms with a model-based approach.

7.4.1 Actor–critic methods

A policy represented by a neural network can be updated by gradient ascent for both the deterministic and the stochastic case. In both cases, the policy gradient typically requires an estimate of a value function for the current policy. One common approach is to use an actor–critic architecture that consists of two parts: an actor and a critic. The actor refers to the policy and the critic to the estimate of a value function (e.g., the Q-value function). In deep RL, both the actor and the critic can be represented by nonlinear neural-network function approximators. The actor uses gradients derived from the policy gradient theorem and adjusts the policy parameters w. The critic, parameterized by θ, estimates the approximate value function for the current policy π: $Q(s, a; \theta) \approx Q^\pi(s, a)$.

Critic: From a (set of) tuples $\langle s, a, r, s' \rangle$, possibly taken from a replay memory, the simplest off-policy approach to estimating the critic is to use a pure bootstrapping algorithm $TD(0)$ where, at every iteration, the current value $Q(s, a; \theta)$ is updated towards the target value

$$Y_k^Q = r + \gamma Q\left(s', a = \pi\left(s'\right); \theta\right). \tag{7.21}$$

The ideal is to have an architecture that is:

- sample-efficient such that it should be able to make use of both off-policy and on-policy trajectories (i.e., it should be able to use a replay memory), and
- computationally efficient: it should be able to profit from the stability and the fast reward propagation of on-policy methods for samples collected from near on-policy behavior policies. There are many methods that combine on- and off-policy data for policy evaluation.

Actor: From Eq. (7.17), the off-policy gradient in the policy improvement phase for the stochastic case is given as

$$\nabla_w V^{\pi_w}(s_0) = \mathbb{E}_{s \sim \rho^{\pi_\beta}, a \sim \pi_\beta}\left[\nabla_\theta\left(\log \pi_w(s, a)\right) Q^{\pi_w}(s, a)\right], \tag{7.22}$$

where β is a behavior policy generally different than π, which makes the gradient generally biased.

7.5 Case study 1: traffic signal control

The typical approach that transportation researchers take is to cast traffic signal control as an optimization problem under certain assumptions about the traffic model, e.g., vehicles come at a uniform and constant rate. Various assumptions have to be made in order to make the optimization problem tractable. These assumptions, however, usually deviate from the real world, where the traffic condition is affected by many factors such as driver's preference, interactions with vulnerable road users (e.g., pedestrians, cyclists, etc.), weather, and road conditions. These factors can hardly be fully described in a traffic model. Thus, researchers may solve the traffic signal control problem in a simplified scenario, as depicted in Fig. 7.2. The left subfigure shows a simplified traffic network with grid-based traffic intersections, and the right subfigure demonstrates a simplified intersection.

On the other hand, reinforcement learning methods can directly learn from the observed data without making unrealistic assumptions about the traffic model, by first taking actions to change the signal plans and then learning from the outcomes. In essence, an RL-based traffic signal control system observes the

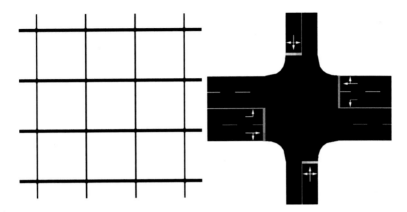

FIGURE 7.2 Traffic signal control.

traffic condition first, then generates and executes different actions (i.e., traffic signal plans). It will then learn and adjust the strategies based on the feedback from the environment. However, in traditional RL-based methods, the states in an environment are required to be discretized and low-dimensional, which is one of the major limitations of the traditional approaches.

Recent advances in RL, especially deep RL, offer the opportunity to efficiently work with high-dimensional input data (like images), where the agent can learn a state abstraction and a policy approximation directly from its input states. A series of related studies using deep RL for traffic signal control have appeared in the past few years.

7.5.1 Agent formulation

A key question for RL is how to formulate the RL agent, i.e., the reward, state, and action definition. In this subsection, we introduce the advances in the reward, state, and action design in recently developed deep RL-based methods.

- **Reward**: The choice of reward reflects the learning objective of an RL agent. In the traffic signal control problem, although the ultimate objective is to minimize the travel time of all vehicles, travel time is hard to serve as a valid reward in RL. Because the travel time of a vehicle is affected by multiple actions from traffic signals and vehicle movements, the travel time as a reward would be delayed and ineffective in indicating the goodness of the signals' action. Therefore, the existing literature often uses a surrogate reward that can be effectively measured after an action, considering factors like average queue length, average waiting time, and average speed or throughput. Researchers also take the frequency of signal changing and the number of emergency stops into consideration for the reward. With various reward functions being proposed, according to (Wei et al., 2021) researchers find out that the weight of each factor in the reward is tricky to set, and a minor difference in weight setting could lead to dramatically different results. Thus, they set out to find a minimal set of factors, proving that using queue length as the reward for a single intersection scenario and using pressure, a variant of queue length in the multi-intersection scenario, are equivalent to optimizing the global travel time.

- **State**: At each time step, the agent receives some quantitative descriptions of the environment as the state to decide its action. Various kinds of elements have been proposed to describe the environment's state, such as queue length, waiting time, speed, phase, etc. These elements can be defined on the lane level or road segment level, and then concatenated as a vector. In earlier work using RL for traffic signal control, people need to discretize the state space and use a simple tabular or linear model to approximate the state functions for efficiency. However, the real-world state space is usually huge, which confines the traditional RL methods in terms of memory or performance. With advances in deep learning, deep RL methods are proposed to handle large state space as an effective function approximator. Recent studies propose to use images to represent the state, where the position of vehicles are extracted as an image representation. With variant information used in state representation in various studies, Wei et al. (2021) also mentioned that complex state definition and large state space do not necessarily lead to significant performance gain and proposes to use simple state like lane-level queue length and phase to represent the environment state.
- **Action**: Now there are different types of action definitions for an RL agent in traffic signal control: (1) set current phase duration; (2) set the ratio of the phase duration over predefined total cycle duration; (3) change to the next phase in predefined cyclic phase sequence; (4) choose the phase to change to among a set of phases. The choice of action scheme is closely related to specific settings of traffic signals. For example, if the phase sequence is required to be cyclic, then the first three action schemes should be considered, while "choosing the phase to change to among a set of phases" can generate flexible phase sequences.

7.6 Case study 2: car following control

Autonomous driving technology is capable of providing convenient and safe driving by avoiding crashes caused by driver errors. Considering, however, that we will likely be confronting a several-decade-long transition period when autonomous vehicles share the roadway with human-driven vehicles, it is critical to ensure that autonomous vehicles interact with human-driven vehicles safely. Because driving uniformity is a large factor in safety, a significant challenge of autonomous driving is to imitate, while staying within safety bounds, human driving styles, i.e., to achieve human-like driving. Human-like driving will: 1) provide passengers with comfortable riding, and confidence that the car can drive independently; and 2) enable surrounding drivers to better understand and predict autonomous vehicles' behavior so that they can interact with it naturally. To achieve human-like driving, it is useful to introduce a driver model that reproduces individual drivers' behaviors and trajectories. Replicating driving trajectories is one of the primary objectives of modeling vehicles' longitudinal motion in traffic flow. Known as car-following models, these models are essential components of microscopic traffic simulation, and serve as theoretical references for autonomous car-following systems.

To address these limitations, a model is proposed for the planning of human-like autonomous car following that applies deep reinforcement learning (deep RL). Deep RL, which combines reinforcement learning algorithms with deep neural networks to create agents that can act intelligently in complex situations, has witnessed exciting breakthroughs such as deep Q-network. Deep RL has the promise to address traditional car-following models' limitations in that: 1) deep neural networks are adept at general purpose function approximators that can achieve higher accuracy in approximating the complicated relationship between stimulus and reaction during car following; 2) reinforcement learning can achieve better generalization capability because the agent learns decision-making mechanisms from training data rather than parameter estimation through fitting the data; and 3) by continuously learning from historical driving data, deep RL can enable a car to move appropriately in accordance with the behavioral features of its regular drivers.

Fig. 7.3 shows a schematic diagram of a human-like car-following framework based on deep RL. During the manual driving phase, data are collected and stored in the database as historical driving data. These data are then fed into a simulation environment where an RL agent learns from trial and error interactions by the environment including a reward function that signals how much the agent deviates from the empirical data. Through these interactions, an optimal policy, or car-following model is developed that maps in human-like behavior from speed, relative speed between a lead and following vehicle, and inter-vehicle spacing to the acceleration of a following vehicle. The model, or policy, can be continuously updated when more data are fed in. This optimal policy will act as the executing policy in the autonomous driving phase.

7.6.1 Agent formulation

Reinforcement learning (RL) aims to address problems related to sequential decision- making by setting up simulations in which an agent interacts with an environment over time. At time step t, an RL agent observes a state s_t and chooses an action a_t from an action space A, according to a policy $\pi (a_t \mid S_t)$ that maps from states to actions. The agent gets a reward r_t for the action choice and moves to the next state s_{t+1}, following the environmental dynamic, or model. When the agent reaches a terminal state, this process stops and then restarts. The task is to maximize the expectation of a discounted, accumulated reward

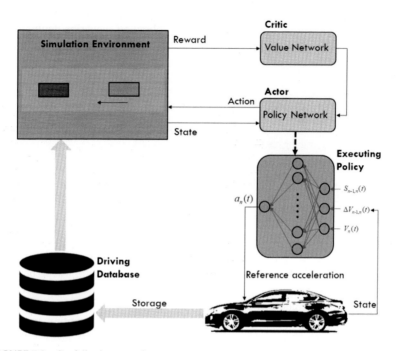

FIGURE 7.3 Car following control.

$R_t = \sum_{k=0}^{\infty} \gamma^k r_{t+k}$ from each state, where $\gamma \in (0, 1]$ represents the discount factor.

Specifically, in the modeling of car-following behavior, the state at a certain time step t is described by the following key parameters: the speed of a following vehicle $V_n(t)$, inter-vehicle spacing $S_{n-1,n}(t)$, and relative speed between a lead and following vehicle $\Delta V_{n-1,n}(t)$. The action is the longitudinal acceleration of the following vehicle $a_n(t)$, confined between -3 m/s^2 and 3 m/s^2 based on the acceleration of all observed car-following events. The reward value is given based on deviations between predicted and observed car-following trajectories. With state and action at time step t, a kinematic point-mass model was used for state updating:

$$V_n(t+1) = V_n(t) + a_n(t) \cdot \Delta T$$
$$\Delta V_{n-1,n}(t+1) = V_{n-1}(t+1) - V_n(t+1)$$
$$S_{n-1,n}(t+1) = S_{n-1,n}(t) + \frac{\Delta V_{n-1,n}(t) + \Delta V_{n-1,n}(t+1)}{2} \cdot \Delta T, \tag{7.23}$$

where ΔT is the simulation time interval, set as 0.1 s in this study, and V_{n-1} is the velocity of lead vehicle, which was externally inputted.

This case study used the reward function $r(s, a)$ to facilitate human-like driving by minimizing the disparity between the values of simulated and observed behavior. Based on these special car-following scenarios, both spacing and speed (velocity) were tested to investigate the difference depending on the type of reward function:

$$r_t = \log \left(\left| \frac{S_{n-1,n}(t) - S_{n-1,n}^{obs}(t)}{S_{n-1,n}^{obs}(t)} \right| \right)$$

$$r_t = \log \left(\left| \frac{V_n(t) - V_n^{obs}(t)}{V_n^{obs}(t)} \right| \right),$$

(7.24)

where $S_{n-1,n}(t)$ and $V_n(t)$ are the simulated spacing and speed, respectively, in the RL environment at time step t, and S_t^{obs} and $V_n^{obs}(t)$ are the observed spacing and speed in the empirical data set at time step t.

7.6.2 Model and simulation settings

In this case study, the DDPG approach to modeling the car-following behavior is introduced as shown in Fig. 7.3. Two separate neural networks were used to represent the actor and critic. At time step t, the actor network takes a state $s_t = (v_n(t), \Delta v_{n-1,n}(t), \Delta S_{n-1,n}(t))$ as input, and outputs a continuous action: the following vehicle's acceleration $a_n(t)$. The critic network takes a state st and an action a_t as input, and outputs a scalar Q-value $Q(s_t, a_t)$. In this case study, the actor and critic networks each have three layers. The final output layer of the actor network used a *tanh* activation function, which maps a real-valued number to the range $[-1, 1]$, and thus can be used to bound the output actions.

Simulation Settings: Driving data collected in the Shanghai Naturalistic Driving Study (SH-NDS). A simple car-following simulation environment was implemented to enable the RL agent to interact with the environment through a sequence of states, actions, and rewards. Data from the lead vehicle served as externally controlled input. The following vehicle's speed, spacing, and velocity differences were initialized with the empirical SH-NDS data. After the acceleration was computed by the agent, future states of the following vehicle were generated iteratively, based on the state-updating rules defined in Eq. (7.23). The simulated spacings, or gaps, were compared to the empirical spacing from the SH-NDS data to calculate reward values and simulation errors. The state was re-initialized with the empirical dataset when the simulated car-following event terminated at its maximum time step.

7.7 Case study 3: bus bunching control

In public transport systems, bus bunching refers to the phenomenon where a group of two or more buses arrives at the same bus stop at the same time. Bus bunching has been a long-standing operational problem in urban public transport systems, and it is a primary issue that concerns transit users and affects our perception of service reliability and service efficiency. As first reported by Newell

and Potts (1964), bus services are born unstable and they are by nature susceptible to bus bunching due to the inherent uncertainties in service operation. On
the one hand, the uncertainty in the number of boarding/alighting passengers
increases the variability of bus dwell time. On the other hand, the time-varying
urban traffic conditions (e.g., congestion and signal control) and the differences
in bus driving behavior also introduce considerable uncertainty in travel time. In
real-world operations, a tiny disruption could result in multiple buses falling behind schedule. When this happens, the delayed buses in general will encounter
more waiting passengers accumulated during the waiting period, which in turn
will result in longer dwell times and further delays. Bus bunching has several
adverse effects on public transport operation. As bus bunching comes into being,
first, it results in increased headway variability, and thus for transit users, both
waiting time and travel time become longer; second, as the system operates, the
buses tend to have imbalanced occupancy rates because most passengers prefer
to board the first arriving bus, in particular when passengers do not have any
information about the arrival of the next bus. Therefore, bus bunching eventu-
ally causes ineffective use of the supply of public transport services. In the long
term, bus bunching will also affect users' perception of service reliability and in
turn discourage the use of public transport which is a key solution to sustainable
transportation.

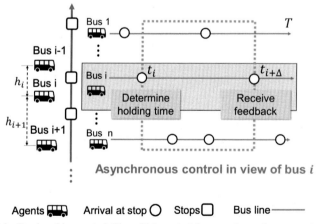

FIGURE 7.4   Bus bunching control. The vehicle control framework on a bus route. A circle indicates the event of a bus arriving at a bus stop. For bus b_i at time t_i, we denote by h_{i+1} the backward
headway (time for the following bus b_{i+1} to reach the current location of b_i) and h_i the forward
headway (time for b_i to reach the current location of b_{i-1}). (*Image source: Wang and Sun, 2021*)

Fig. 7.4 shows the vehicle holding control framework on a bus route, in
which we take all the buses running in the system/environment as agents. Note
that we label all buses in a sequential order, where bus b_{i+1} follows bus b_i. Bus
holding control for a fleet $\{b_i\}$ is a typical application of asynchronous multi-

agent control, as holding control is applied only when a bus arrives at a bus stop.

7.7.1 Agent formulation

We define the basic elements in this framework as follows:

- **State**: Each agent has a local observation of the system. For bus b_i arriving at stop k_j at time t, we denote the state of b_i by $s_{i,t}$, which includes the number of passengers onboard, the forward headway h_i, and the backward headway h_{i+1}.
- **Action**: Following (Wang and Sun, 2021), we model holding time as $\Delta d_{i,t} = a_{i,t} \Delta T$, where ΔT is the maximum holding duration and $a_{i,t} \in [0, 1]$ is a strength parameter. We consider $a_{i,t}$ the action of bus b_i when arriving at a bus stop at time t. Here, ΔT is used to limit the maximum holding duration and avoid over-intervention. We model $a_{i,t}$ as a continuous variable in order to explore the near-optimal policy in an adequate action space. Note that no holding control is implemented when $a_{i,t} = 0$.
- **Reward**: Despite that holding control can reduce the variance of bus headway and promote system stability, the slack/holding time will also impose additional penalties on both passengers (i.e., increasing travel time) and operators (i.e., increasing service operation hours). To balance system stability and operation efficiency, we design the reward function associated with bus b_i at time t as

$$r_i^t = -(1 - w) \times CV^2 - w \times a_{i,t}, \qquad (7.25)$$

where $CV^2 = \frac{\text{Var}[h]}{E^2[h]}$ quantifies headway variability and $w \in [0, 1]$ is a weight parameter. Essentially, $E[h]$ is almost a constant given the schedule/timetable, and thus CV is mainly determined by $\text{Var}[h]$: a small CV indicates consistent headway values on the bus route, and a large CV suggests heavy bus bunching. The second term in Eq. (7.25) penalizes holding duration and prevents the learning algorithm from making excessive control decisions. This term is introduced because any holding decisions will introduce additional slack time and reduce operation efficiency. Overall, the goal of this reward function is to effectively achieve system stability with as few interventions as possible, with w serving as a balancing parameter.

The agent implements an action (i.e., determining holding time) when it arrives at a stop, and then it will receive feedback upon arriving at the next stop. This dynamic process is considered a multi-agent extension of Markov decision processes (MDPs) and referred to as a Markov game $G = (N, S, A, P, R, \gamma)$, where N denotes the number of agents, γ denotes the discount factor, S and $A = \{A_1, \ldots, A_N\}$ denote the state space and the joint action space, respectively, and $P : S \times A \mapsto S$ represents the state transition function. However, this definition becomes inaccurate in the bus control problem, since agents rarely implement actions at the same time. To address this asynchronous issue, we introduce a modified formulation in which each agent maintains its own transition: $\hat{\mathcal{P}}_i : \mathcal{S}_i \times \hat{\mathcal{A}} \mapsto \mathcal{S}_i, \hat{\mathcal{A}} \subseteq \mathcal{A}$, where $\mathcal{S}_i$ is the observation of agent i. In this way, state transition observed by agent i does not necessarily depend on actions from all the other agents at any particular time. The policy to choose an action is given by the conditional probability $\pi_{\theta_i} = p(A_i \mid S_i)$. Finally, the reward function for each agent can also be defined independently: $\mathcal{R} = \{R_1, \ldots, R_N\}$.

These examples show RL can be applied to many control problems in the transportation domain.

7.8 Exercises

7.8.1 Questions

1. What's the difference between Q-learning and policy gradients methods?
2. What's the difference between a deterministic and stochastic policy?
3. Why would you use a policy-based method instead of a value-based method?
4. What is the difference between off-policy and on-policy learning?
5. What is the difference between the discount factor $\gamma = 0$ and $\gamma = 1$ in reinforcement learning?
6. When applying RL to the single-intersection traffic signal-control application, which of the following cannot be the state S?
   a. Queue length on each lane of the road;
   b. Number of vehicle on each lane of the road;
   c. Traffic signal phase;
   d. Average traffic speed of each lane.
7. What is the specific advantage of DQN?
8. List three methods to incorporate the communication between the agents when you use multi-agent reinforcement learning to solve the multi-intersection traffic signal control problem.
9. What are the parameters that you can tune when training neural network-based RL models?

Chapter 8

Transfer learning

8.1 What is transfer learning

In the classic supervised learning scenario of machine learning, if we intend to train a model for some task and domain A, we assume that we are provided with labeled data for the same task and domain. For the moment, let us assume that a task is the objective our model aims to perform, e.g. recognize objects in images, and a domain is where our data is coming from, e.g., images taken in Seattle coffee shops. We can now train a model A on this dataset and expect it to perform well on unseen data of the same task and domain. On another occasion, when given data for some other task or domain B, we require again labeled data of the same task or domain that we can use to train a new model B so that we can expect it to perform well on this data. The traditional supervised learning paradigm breaks down when we do not have sufficient labeled data for the task or domain we care about to train a reliable model.

If we want to train a model to detect pedestrians on night-time images, we could apply a model that has been trained on a similar domain, e.g. on day-time images. In practice, however, we often experience deterioration or collapse in performance as the model has inherited the bias of its training data and does not know how to generalize to the new domain. If we want to train a model to perform a new task, such as detecting bicyclists, we cannot even reuse an existing model because the labels between the tasks differ. Transfer learning allows us to deal with these scenarios by leveraging the already existing labeled data of some related task or domain. We try to store this knowledge gained in solving the source task in the source domain and apply it to our problem of interest. In practice, we seek to transfer as much knowledge as we can from the source setting to our target task or domain. This knowledge can take on various forms depending on the data: It can pertain to how objects are composed to allow us to more easily identify novel objects; it can be with regard to the general words people use to express their opinions; etc.

8.2 Why transfer learning

It is indisputable that ML use and success in industry has so far been mostly driven by supervised learning. Fueled by advances in deep learning, more capable computing utilities, and large labeled datasets, supervised learning has been largely responsible for the wave of renewed interest in AI, funding rounds and acquisitions, and, in particular, the applications of machine learning that we have seen in recent years and that have become part of our daily lives. Over recent years, we have been able to train more and more accurate ML models using supervised learning. We are now at the stage that the state-of-the-art models have achieved super-human ability in many ML tasks, such as object recognition, speech generation, and disease identification, among many others.

However, these successes rely heavily on labeled data to reach the current level of performance. These data may have been collected over years. In a few cases, it is public, e.g., ImageNet, but large amounts of labeled data are often proprietary or expensive to acquire. At the same time, when applying a machine learning model in the real-world, we are faced with a myriad of conditions that the model has never seen before and does not know how to deal with; each client and every user has their own preferences and possesses or generates data that is different than the data used for training; a model is asked to perform many tasks that are related to, but not the same as, the task it was trained for. In all of these situations, our current state-of-the-art models, despite exhibiting human-level or even super-human performance on the task and domain they were trained on, suffer a significant loss in performance or even break down completely.

Transfer learning can help us deal with these novel scenarios and is necessary for production-scale use of machine learning that goes beyond tasks and domains where labeled data is plentiful. So far, we have applied our models to the tasks and domains that—while impactful—are the low-hanging fruits in terms of data availability. To also serve the long tail of the distribution, we must learn to transfer the knowledge we have acquired to new tasks and domains.

8.3 Definition

The definition closely follows the popular survey paper Pan and Yang (2009), in which binary document classification is a running example. Transfer learning includes a domain and a task in the concept. A domain D contains a feature space F and a marginal probability distribution $P(X)$ over the feature space, where $X = x_1, ..., x_n \in F$. For document classification with a bag-of-words representation, F is the space of all document representations, x_i is the i-th term vector corresponding to some document, and X is the sample of documents used for training.

Given a domain, $D = F, P(X)$, a task T consists of a label space y and a conditional probability distribution $P(Y|X)$ that is typically learned from the training data consisting of pairs $x_i \in X$ and $y_i \in y$. In the document classification example, y is the set of all labels, and $y_i = 0$ or 1.

Given a source domain D_S, a corresponding source task T_S, as well as a target domain D_T and a target task T_T, the objective of transfer learning is to enable the learning of the target conditional probability distribution $P(Y_T|X_T)$ in D_T with the information gained from D_S and T_S, where $D_S \neq D_T$ or $T_S \neq T_T$. In most cases, a limited number of labeled target examples, which is exponentially smaller than the number of labeled source examples, are assumed to be available.

Given source and target domains D_S and D_T, where $D = F, P(X)$, and source and target tasks T_S and T_T, where $T = y, P(Y|X)$ source and target conditions can be divided into four types. The first type is $F_S \neq F_T$. The feature spaces of the source and target domain are different. The second type is that $P(X_S) \neq P(X_T)$, i.e., the marginal probability distributions of source and target domain are different, which is normally called domain adaptation. The third type is $y_S \neq y_T$, where the label spaces between the two tasks are different. The fourth type of transfer learning is $P(Y_S|X_S) \neq P(Y_T|X_T)$, where the conditional probability distributions of the source and target tasks are different.

8.4 Transfer learning steps

In a general setting of adopting deep neural networks, transfer learning generally works in four steps:

- Pretrained model acquisition;
- Freezing the layers;
- Processing new layers;
- Model fine-tuning.

First and foremost, depending on the tasks, a pretrained model is acquired. This model is usually pretrained on a more generalized task. There should be strong correlations between the knowledge of the source domain and that of the target domain to make them compatible. The pretrained model usually has an architecture of well-known neural networks, such as VGG, ResNet, and Inception. Pretrained neural network weights can often be downloaded. Conditions like training datasets, parameter settings, and downstream tasks have impact on the pretrained model performances. In the transfer learning process, it is critical to freeze the initial layers in order to save training time by avoiding the learning of simple features. Without layer freezing, there would be no difference from training a neural network from scratch. There are two cases in the structure of the following layers: with our without adding new trainable layers. In either case, we reuse the knowledge from the basic features and update the high-level features in the neural network. The case of adding new layers is for the specialized tasks of the new network. For example, it is for the same task of road users classification. However, the original network is trained to have three output classes: light vehicles, heavy vehicles, and nonmotorized users. The updated task is more specific and needs to classify the road users into more classes, e.g., single-unit truck, van, motorcycle, pedestrian, sedan. The unfrozen layers will be retrained, and any new layers will be trained with the new data. Lastly, as in normal deep learning model training, the models are usually fine-tuned to further improve the performance. An example of a fine-tuning method is to unfreeze some part of the neural network again (could include the first few layers) and then retrain at a minimal learning rate or with a specific focus on some subclasses.

8.5 Transfer learning types

8.5.1 Domain adaptation

In domain adaptation, the source and target domains have different feature spaces and distributions. It is the process of adapting one or more source domains to transfer information to improve the performance of a target learner. This process attempts to alter a source domain to bring the distribution of the source closer to that of the target. The mechanism of domain adaptation is to uncover the common latent factors across the source and target domains, and then use the latent factors to minimize the mismatch regarding feature spaces between the two domains. Another two key terms in domain adaptation are domain translation and domain shift, where domain translation is a technique that essentially finds the correspondence between the source domain and target domain, and domain shift is a change in the data distribution between different domains.

8.5.2 Multi-task learning

In the case of multitask learning, several tasks from the same domain are co-learned simultaneously without distinction between the source and targets. We have a set of learning tasks and we co-learn all tasks simultaneously. This helps to transfer knowledge from each scenario and develop a rich combined feature vector from all the varied scenarios of the same domain. The learner optimizes the learning/performance across all of the n tasks through some shared knowledge. When it comes to learning a large number of tasks with a meta-learner, it is called meta-learning. The primary goal of meta-learning is not to learn a single classifier using supervised learning but a model that adapts to new tasks efficiently. The deep neural network model normally embeds data in a common representation feature space. The process is repeated over a distribution of tasks, where each given task is a subtask of the target problem.

8.5.3 Zero-shot learning

Data labeling is a labor-intensive job: Obtaining annotations needs time and sometimes domain expertise, where specialized experts are needed; moreover, data imbalance in the real world could be a problem, since it is impossible to enumerate all the cases to be "seen" in a training dataset. Classifying objects into unseen classes, without relying on labeled data samples, is a critical requirement in current and future autonomous systems.

Zero-shot learning is a strategy where the pretrained learning model is generalized to a novel category. The training and testing distributions are disjoint, but at the same time, zero-shot learning needs additional data during the training phase to better understand the unseen data distribution. For instance, a model is trained using a traffic surveillance image dataset for road user recognition, and no emergency vehicles (e.g., ambulances, fire trucks) are included in the data samples, but this model is now made to recognize emergency vehicles in surveillance images. The road user classes covered in the training samples are the "seen" classes, while the emergency vehicles unlabeled in the test samples are referred to as the "unseen" classes.

Zero-shot learning is a subfield of transfer learning in a way that it transfers the knowledge already contained in the training samples to the testing tasks. A straightforward example is that of road user recognition: Features for trucks are learned and, if combined with the color information and features of stairs from another object recognition model, it already knows the individual features to identify a fire truck, which is a red truck in with stairs. When a transfer learning problem is with the same feature and label spaces, it is called homogeneous transfer learning. Zero-shot learning belongs to the category of heterogeneous transfer learning where the feature and label spaces are disparate.

In zero-shot learning, the data contains seen classes, unseen classes, and auxiliary information. Seen classes are the data that usually have clear and handcrafted labels, which are used to train the machine learning model; unseen classes are the data without explicit labels that the existing learning model will generalize to cover; auxiliary information is the bridge between the seen classes and the unseen classes, which is necessary for the generalization step since no explicit labels are available for the unseen categories. The auxiliary information needs to cover all the unseen classes, and can be in various formats, e.g., text descriptions or semantic information, depending on the zero-shot learning scheme.

Just like traditional machine learning, zero-shot learning has a training stage and an inference stage. We use image classification as an example since it is still the most common zero-shot learning task. In the training stage of the method, it learns about images and their corresponding auxiliary information. If the image is a cat, and you can design a semantic encoding attribute vector as the auxiliary information, which contains "tail, beak, feather, whiskers, and furry." For a cat, the tail, whiskers, and furry elements are labeled as 1 and the beak and feather are 0. Therefore, the collection of seen data classes, seen labels, and the attribute vector form the seen set. There is also a set of data classes as unseen classes, which can also be described using the same representation, with unseen images, unseen labels, and attribute vectors. In the animal example, zebra could be an unseen class in the inference stage, where all the attributes needed to identify a zebra have been seen (e.g., stripes from tigers, the shape of horses, color from pandas). Another representative way of doing zero-shot learning could be that the model learns about images by "reading" auxiliary text descriptions (e.g., sentences) associated with each image. It could be "there are two cute dogs on the sofa." A human can easily interpret and decipher this sentence that the two similar things in the image are called dogs. By seeing millions of image–text pairs of different kinds of animals in this case, the learning model accumulates knowledge to extrapolate to other image classification tasks. A text encoder and an image encoder for each pair are generated and will be used for the zero-shot prediction of unseen images.

8.5.4 Few-shot learning

Humans can identify the characteristics of a class of objects after seeing a few examples. However, computers normally need a large amount of data to learn the features and identify what they see. Few-shot learning trains computers to learn from a few samples like humans. Few-shot learning is expected to benefit rare cases and unbalanced data training. For instance, corner cases in automated driving are rare cases that are not commonly seen but could cause unexpected consequences. Few-shot learning has the potential to address this issue by correctly detecting these corner case events by exposing them to only a few. Similar to zero-shot learning, few-shot learning can also reduce labor costs and computation power. Few-shot learning is different from zero-shot learning in a way that few-shot learning predicts the classes where a small number of data samples are available, while zero-shot learning assumes the class is completely unseen.

The primary goal of few-shot learning is to learn a similarity function that maps the similarities between the classes in the support set and query set. The support set contains a few labeled samples for the new data class. The query set contains the samples from both novel and old data classes on which the model needs to generalize using prior knowledge learned from the support set. A common learning scheme used in few-shot learning is called the N-way K-shot learning scheme. It basically describes the few-shot problem statement that a model handles. N-way means that there are N numbers of new classes on which a learning model needs to generalize. K-shot indicates the number of labeled data samples that are available in the support set for each of the new classes. The larger the N value and the smaller the K value, the task becomes more difficult. K is typically in the range 1–5, where K=1 has the name "one-shot learning" and, if K=0, it becomes "zero-shot learning," which we previously discussed, and the solution approaches are very different from few-shot learning.

8.6 Case study: vehicle detection enhancement through transfer learning

In a practical application, the authors apply Transfer Learning (TL) to enhance the vehicle detection performance in the real-world on an IoT device, i.e., Raspberry Pi, for the purpose of parking occupancy detection (Ke et al., 2020d). An SSD-Mobilenet detector is implemented using TensorFlow Lite on the IoT devices with TL on the MIO-TCD traffic surveillance dataset.

SSD with a Mobilenet backbone network is the primary detector. There are multiple backbones for SSD, while Mobilenet has the lightest structure which makes the detection faster than other backbones. This is appropriate for an IoT device with limited computational power. We recommend using TensorFlow Lite for the SSD implementation since it is designed for deep learning on mobile and IoT devices. A normal state-of-the-art object detector like YOLO with the TensorFlow platform still runs slowly with a speed lower than 0.05 frames-per-second (FPS) on a Raspberry Pi 3B, and has a slightly lower detection accuracy as well. However, SSD-Mobilenet with TensorFlow Lite runs over 1 FPS on the same device according to our test. The detection results including bounding boxes, object type, and detection probabilities (how likely the result is true) are transmitted back to the server. Compared to sending videos, it reduces the data volume by thousands of times (the exact number depends on the number of detections in the video).

TensorFlow models can be converted to TensorFlow Lite models. We recommend training a TensorFlow model and then converting it into the TensorFlow Lite model. In order to improve the detection performance to make it more appropriate for practical applications, we enhance a pretrained SSD on the Pascal VOC dataset (Everingham et al., 2010) with a new traffic surveillance dataset called MIO-TCD (Luo et al., 2018), which contains 110,000 surveillance camera frames for traffic object-detection training. This dataset includes a variety of challenging scenarios for traffic detection such as nighttime, truncated vehicle, low resolution, shadow, etc. To our knowledge, this is the first time MIO-TCD been adopted for parking detection, and we find it works well.

Some key parameters for the training are listed as follows: the learning rate is 0.00001, the weight decay is 0.0005, the optimizer is Adam, the batch size is 32, and the training-validation split ratio is 10:1. All layers are trainable. The training and validation loss curves, as well as some sample images at certain training steps, are displayed in Fig. 8.1.

The enhanced SSD-Mobilenet model demonstrates great performances on traffic detection, especially in challenging surveillance image data. Fig. 8.2 shows three examples comparing detection results between SSD trained on Pascal VOC and Pascal VOC + MIO-TCD. In the first column, the pretrained SSD detects all big targets but misses two small targets in the back; in the second column, the pretrained SSD misses two vehicles partially blocked by a tree; in the third column where there is snow in the nighttime, the pretrained SSD misses most of the vehicles. Overall, the enhanced SSD produces much better detection results with few missed detections and no false detections.

8.7 Case study: parking information management and prediction system by attribute representation learning

8.7.1 Background

The case study from the authors (Yang et al., 2022) is introduced in detail with a focus on attribute representation learning. Representation learning, also called

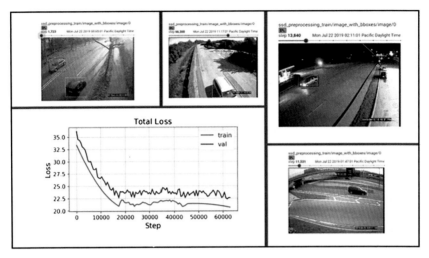

FIGURE 8.1 The sample images from MIO-TCD dataset and the transfer learning curves.

FIGURE 8.2 Example vehicle detection improvements before and after transfer learning.

feature learning, is the learning task in which neural networks try to determine which feature to use.

Parking shortages and challenges are not only very serious for cars in the city region, but also for trucks. Although not reflected by parking research, the imbalance of supply and demand for truck parking has become a common concern (Association et al., 2017). Truck-borne freight is a key component of the modern supply chain system all over the world. Taking the USA as an example, in 2015, the total tonnage of trucks was 10.49 billion tons, accounting for 70.1% of the total tonnage of the domestic freight industry (Association et al., 2017; Wrenn, 2017). Among the transport, 30%–40% are long-distance shipments (more than 750 miles) with over 50% (Sprung et al., 2018) cargo values. In order to complete such long-distance driving tasks safely and efficiently, truck drivers need adequate and high-quality rest. No doubt, timely truck parking information would be necessary and valuable for them.

To solve the ubiquitous parking challenges for cars and trucks, the first and foremost approach is to make full use of existing parking resources instead of building new ones. In this case, disseminating real-time and future parking availability information to drivers would significantly help them schedule their stops at parking facilities. Therefore, the Parking Information Management System (PIMS) has been gradually used to improve the parking efficiency and utility (Yan et al., 2011; Wang and He, 2011; Geng and Cassandras, 2013; Cheng et al., 2020b; Bayraktar et al., 2015; Sun et al., 2018). PIMS involves typically three major components: parking detection, data processing, and information dissemination (Geng and Cassandras, 2013; Lin et al., 2017; Sun et al., 2018). First, real-time parking data is collected by parking sensor technologies. Then, the sensor data is processed to calculate the current parking availability. Finally, the real-time parking utility information is disseminated to drivers via approaches such as dynamic message signs, websites, mobile apps, etc.

With the deployment of PIMS, high-quality parking data can be collected. Then, data-driven methodologies for solving parking issues are being developed and investigated by researchers.However, in real life, there are so many factors showing impacts on the parking utility. The complex features sets can be summarized into three categories, the spatiotemporal set (Sadek et al., 2020), the attributes information set (i.e., weather impact, hour of a day, and day of a week) (Haque et al., 2017), and the driver's behavior set (i.e., parking duration information, parking habits, etc.) (Vital et al., 2020). The integration of these factors with heterogeneous formats into traditional prediction methods becomes a challenging issue.

The cutting-edge learning-based models perform well in certain scenarios. However, their limitations are also obvious, so several key challenges need to be overcome. The first and foremost one is the attributes information integration. Previous studies indicated that the attribute information has a significant impact on the parking pattern (Yang et al., 2019; Liu et al., 2014; Zhang et al., 2020a; Sun et al., 2018). However, in contrast to the spatiotemporal record, the attributes information always has heterogeneous values with different formats, including category values, clusters, and even text information. If using these features directly as the network input, these characteristics' impacts will be greatly limited and even trigger negative impacts (Frank et al., 2001; Hoi et al., 2006, 2010). Besides successfully integrated the attributes features into training process, balancing attributes and historical impacts and fusing features to a reliable prediction result need to be investigated. An attributes-aware attention mechanism is necessary to be integrated. The second challenge is the limited scalability issues. Usually, neural networks are designed with a particular motivation, especially in parking problems. It is difficult for different sensor systems to share the same predictive model, which is especially obvious in deep learning approaches. Therefore, a systematic prediction scheme that shares common input and output and can serve all sensor systems (centralized and decentralized) on the market is a necessity. The third hurdle is the limited transferability of the current spatiotemporal neural networks. As is known, the neural network prediction result is trained based on the ground truth data. If a new parking lot with limited parking data needs to be integrated into the predicting framework, the whole model needs to be retrained or even rebuilt, which is very time-consuming and inconvenient. These limitations seriously restrict the field deployment of the neural network-based parking prediction algorithms.

8.7.2 Methods

In this research, the team sought to solve the three main challenges from both the theoretical and practical aspects. Representation learning is introduced and integrated into the model to better capture the attributes features and increase the flexibility of prediction framework. Instead of using original information as input, the goal of representation learning is to learn a representative features and measure the correlations among objects feature sets by transferring the raw data into a high-dimensional vector space (Li et al., 2019b; Chen et al., 2019a). The success of the method is tested and implemented by many computer science topics such as natural language processing (NLP) (Li et al., 2017; Kulis et al., 2012), Computer Vision (CV) (Cakir et al., 2019; Geng et al., 2011; Huang et al., 2019) and Recommendation Systems (RS) (Avazpour et al., 2014; Da'u and Salim, 2019). Thus, with the heterogeneous feature embedding and representation learning, the research team can better capture and combine the spatiotemporal features, attribute features, and the customized information provided by the users (i.e., the arriving time) into back-propagation training process. Such a modeling structure can make the results more universal and accurate for truck parking and urban parking. Instead of using raw sensor data as the neural network input, a general preprocessing workflow is designed using the summarized parking lot-occupancy sequences as the prediction input. With the feature learning process, the transferability of the prediction framework is greatly improved. Also, inspired by the broad learning architecture (Zhang and Philip, 2019; Liu et al., 2020b), the proposed prediction framework takes the future flexibility and applicability into account, which can easily integrate the attributes features with different formats and predict the target based on various external condition inputs, instead of using only spatiotemporal information.

The PIMPS system includes four subcomponents: sensing component, data processing component, attributes information collection component, and prediction component. Detailed flowcarts illustrating the procedures can be found in Fig. 8.3.

The sensing component plays a role as the eyes and ears in the intelligent parking system, which is used to detect the status of the target parking region and then summarize into occupied/unoccupied status. In general, the current parking sensor system can be divided into two types of approaches: decentralized (space-by-space) sensing and centralized (entrance/exit-based) sensing

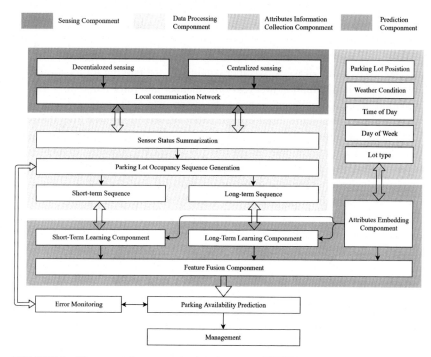

| Sensing Component | Data Processing Component | Attributes Information Collection Component | Prediction Component |

FIGURE 8.3 The proposed representation learning system PIMPS.

architecture. Examples can be found in Fig. 8.2. The decentralized sensing always consists of multiple sensors installed on parking slots to monitor the slot status, each sensor is in charge of one or more slots. The slot-based sensor sensing, i.e., radar senor, magnetic detector, laser detector, and video sensor. Under the circumstance, the decentralized system is more accurate and reliable, but usually has a high cost with a complex installation process since the parking sensors need to be mounted at each parking slot. Another centralized sensing system is more straightforward. The sensors are always installed at the parking lot entry/exit way, and the numbers of driving in and out of the lot can be directly obtained. Compared with the slot-based approach, the entry/exit system is cheap and easy to deploy but may lead to traffic congestion. However, slot level-occupancy information is not feasible. The one-by-one pass of the parking lots sometimes becomes a bottleneck. Also, if the sensing result is not very reliable, the error accumulation may be a big problem when using the entrance/exit count approach.

After the sensing component sends the occupancy status to the local server, the post-processing procedure is started to prepare the input sequences. In general, the decentralized sensing status, as well as the centralized sensing results, are summarized within the overall parking lots occupancy. Then, two sequences, one short-term real-time parking sequence (R_n^i) and long-term historical parking occupancy sequence (H_m^i), are calculated and ready to use.

Parking activity is highly impacted by the status of the attributes including, time of day, day of week, weather conditions, the parking lot location, etc. Collecting such information in a real-time manner and fully integrating the information into the prediction system is necessary. Here, the research team sets up an attributes information collection component served for capturing the necessary external factor, including parking lot type (business, working, shopping, rest area, etc.), weather condition (fair, cloudy, light rain, rain, etc.), time of day, day of week, and type of road (freeway, arterial streets, local streets, etc.), then processed as the input for the prediction component.

For the prediction component, four parts are summarized into the system, including the short-term feature learning component, long-term learning component, the attributes embedding component, and the feature fusion component. The detailed architecture will be discussed in the next section. Then, after the target output sequence is obtained, the error monitoring component is used to watch the predicted result and compare it with the ground truth, and report the comparison result to the parking lot manager. If the error is larger than expected, auto retraining will be activated, and the model will be retrained with the data in the recent eight weeks.

As shown in Fig. 8.4, the PAP neural network is consists of four parts: Input Embed Component (IEC), Attributes Tensor Embedding Component (ATEC), Temporal Learning Component(TLC), and the Feature Fusion Component (FFC).

In order to better enable the network to fully extract the long-term and short-term temporal dependency of parking patterns, in the data input component, the authors use two sets of sequence data as input for the prediction task. The two sets of sequences are real-time sequence (R^i) and history sequence (H^i). The real-time sequence represents the most current parking situation, consisting of n occupancy data records. Meanwhile, the history sequence means at the prediction target time flag, the past parking occupancy record at the same time of a day and day of a week (i.e., if the prediction target of time is next Wednesday, 8:00 AM, for parking lot i, the H^i will be consists of m historical occupancy records captured on the past several Wednesdays, 8:00 AM).

Then, to better extract the hidden pattern of the two sequence for each prediction target, the research team designed a input embedding process for the input sequences. To obtain each prediction target input sequences (In_{T_i}), a matrix embedding

$$In_{H^i} = tanh(W_{H^i} \cdot h_m^i) \tag{8.1}$$

$$In_{R^i} = tanh(W_{R^i} \cdot r_n^i) \tag{8.2}$$

is used to map the m_{th} records of historical and n_{th} real-time occupancy sequence into a high-dimensional vector space $In_{T_i} \in \mathcal{R}^{16}$. And in Eq. (8.1) and

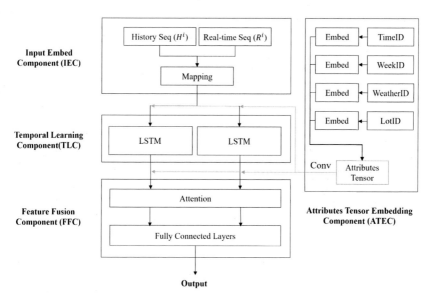

FIGURE 8.4 Overall framework of the parking availability prediction design.

(8.2), the W_{H^i} and W_{R^i} are the learnable embedding weight matrix. Specifically, in this research, the team choose the m equal to n as the same length. So, the output of Eqs. (8.1) and (8.2) In_{T_i} and In_{R^i} are used as the input for occupancy prediction, whose size are $\mathcal{R}^{16*|m|}$.

From the previous research, the attributes information, including the time of day, day of the week, and weather conditions heavily impact the parking occupancy and pattern (Yang et al., 2019; Sun et al., 2018). For example, for some parking lots, the daily packing lot occupancy status of the workday is much higher than that of the weekend. Also, the frequency of short-term parking (less than 20 min) in the midafternoon are much higher than evening and night. Obvious to note, these attributes clusters information have connections and interrelationships with each other. Only vector mapping cannot sufficiently capture the hidden pattern inside of the sets. Furthermore, since the attributes information is always collected as discrete bracket values, directly using them as neural network input is not feasible (Wang et al., 2018; Gal and Ghahramani, 2016). To fully utilize this information, in this research, we design a particular embedding approach, called the Attributes Tensor Embedding Component (ATEC), to extract and integrate the features into the neural network at the utmost level. The ATEC includes two steps, value embedding, and tensor embedding:

$$Att_{a_j^i} = W^{a_j^i} \cdot a_j^i. \tag{8.3}$$

In the equation, the a^i is the attribute information that belongs to each occupancy data record. W^{a^i} is the learnable matrix, whose dimension is $\mathcal{R}^{A*E}$, is used to transform the input attributes category values a^i ($a^i \in A$) from a original space A to the embedding space E. Each attribute information has its own corresponding mapping matrix to ensure the feature can be map into a suitable space.

The second step aims to capture the combination and correlations inside of the attributes vectors. Here, the research team designs a tensor embedding procedure to extract the latent features. The first substep is to build an attribute tensor by using the vector outer product

$$\mathcal{T}_{a^i} = Att_{a_j^1} \otimes Att_{a_j^2} \cdots \otimes Att_{a_j^i}. \tag{8.4}$$

The output of Eq. (8.4) is a tensor $\mathcal{T}_{a^i} \in \mathcal{R}^{E_1 \times E_2 \cdots \times E_j}$. Then, we use a multidimensional convolution operation to extract the attributes tensor features and map into a one-dimensional vector. This substep serves mainly to capture and emphasize the inside correlations of attributes.

$$Att_{TE^i} = \sigma_{cnn}(W_{conv} * \mathcal{T}_{a^i} + b). \tag{8.5}$$

Then, we obtain the output of the attribute set embedding vector Att_{TE^i}, which includes j different factors. Specifically, j is the number of different attributes included inside of the attributes set. We will use the Att_{TE^i} to represent the attributes features in the following components.

After both attribute information and sequence records are transformed into a vector features set, then, the temporal learning component is used to capture the time dependency features. Here, we design two separate stacked LSTM modules to capture both the historical sequence and the real-time occupancy sequences together. The input of the TLC for a prediction target is

$$In_{TLC_{R^i}} = In_{R^i} \circ Att_{TE_{R^i}} \tag{8.6}$$

$$In_{TLC_{H^i}} = In_{T_i} \circ Att_{TE_{H^i}}. \tag{8.7}$$

From these two equations, the input of TLC can be divided into two sequences, and each of them includes the occupancy sequence features and the attributes features. Both features vector is linked by the concatenate operation (∘) and then sent to the temporal learning network. Here, to further capture the dependencies of the input sequence, the team introduced the recurrent neural network (RNN) into our model. RNN is a widely used artificial neural network for learning the temporal features in sequences, especially the transportation data sequence. Generally, RNN includes the following three characteristics:

- RNN can produce an output at each time node, and the connection between hidden units is cyclic;

- RNN can produce an output at each time node, and the output at that time node is only cyclically connected to the hidden unit of the next time node;
- RNN contains hidden units with cyclic connections and can process sequence data and output a single prediction.

In our model, a well-known special RNN–LSTM is designed to capture the truck parking sequence dependency features. The main contribution of the RNN is that it can transfer and learn the features from the previous input to future moments. However, in scenarios that require long-range information dependence, it is very difficult to train a well-performing RNN since the gradient vanishing/exploding problem often appears. The LSTM solves this problem by using special architecture to forget some unimportant information during the training process (Hochreiter and Schmidhuber, 1997).

To capture the temporal features, PAP uses two stacked LSTM to capture the temporal dependencies for the input sequences. The mathematical formulations of weights update for the stacked LSTM layers are shown below:

$$h_i^{H^i} = \sigma_{lstm}(W_x \cdot In_{TLC_{H^i}} + W_h \cdot h_{i-1}^{H^i} + W_a \cdot Att_{TEH^i}) \qquad (8.8)$$

$$h_i^{R^i} = \sigma_{lstm}(W_x \cdot In_{TLC_{R^i}} + W_h \cdot h_{i-1}^{R^i} + W_a \cdot Att_{TER^i}) \qquad (8.9)$$

where $In_{TLC_{R^i}}$, $In_{TLC_{H^i}}$ are the two input data sequences; Att_{TER^i} and Att_{TEH^i} are the attributes features vector after the tensor embedding of each occupancy record; and the W_x, W_h and W_a are learnable parameter matrices used in the LSTM.

Since the research team prefers to use the PAP neural network to integrate heterogeneous latent features, a feature fusion and attention component are necessary. From previous research, the parking utility prediction features can be roughly divided into two parts: short-term features and collective features. Formally, the short-term features mainly serve to predict close-target time-availability information. This information always can be investigated and summarized from the real-time continuous input sequence (i.e., the occupancy sequences or parking status sequences). Meanwhile, when the traveler wants to predict the long-term availability of parking information (much longer than the time gap of the input sequence), the long-term features play a role. Generally, the collective features include the long-term parking pattern features and the attributes conditions. Hence, the model not only needs to refer to the current occupancy sequence, but also the historical occupancy records with similar attribute conditions. Meanwhile, sufficiently integrating the attributes information at the target prediction time is quite necessary since the long-term parking pattern is closely related to the attributes conditions, i.e., day of the week, time of the day, etc.

In PAP, to better fuse and use the two types of features, the FAC is designed and implemented. The first step is the attention mechanism. Here, the research team designs an attention mechanism considering the attributes factors instead of mean pooling. The attention includes three procedures. The first part is the attributes vector similarity calculation. In each prediction target, attributes information includes two parts: the input occupancy sequence (O_i) attributes ($Att_{TE_{R^i}}$, $Att_{TE_{H^i}}$) and the target prediction attributes Att_{TE_T}. The cosine similarity is used here:

$$Sim(T, O_i) = \frac{Att_{TE_T}^{\top} \cdot Att_{TE^{O_i}}}{|Att_{TE_T}||Att_{TE^{O_i}}|}. \tag{8.10}$$

Then, the attention value α_i can be calculated by the following equation:

$$h_i = h_{HAtt} \circ h_{RAtt} \tag{8.11}$$

$$\beta_i = < \sigma_{Att}(Att_{TE^{O_i}} \cdot Sim(T, O_i)), h_i >$$

$$\alpha_i = \frac{e^{\beta_i}}{\sum_j e^{\beta_j}}. \tag{8.12}$$

The $\circ$ is the concatenate operation, and σ_{Att} is the nonlinear mapping that maps the attention into the same dimension with h_i. With the attributes attention, the researchers summarized the two sequence features into an attention-based vectors: h_i. Then, the final features vector ($\mathcal{F}_T$) for the prediction target T is

$$\mathcal{F}_T = \sum_{i=1}^{2m} \alpha_i \cdot h_i. \tag{8.13}$$

Finally, the researchers use four fully connected layers to downsample the $\mathcal{F}_T$ from 64 to only one single neuron, which represents the predicted parking occupancy result (O_{T_i}).

8.7.3 Results

Urban Parking Data: In the research, the team collected 19 parking lots data including 1,534 urban parking slots in Zhengzhou, a midsize city in the central area of China. The parking lots are all in the busy urban area of Zhengzhou (inside the third ring road), and whose locations can be divided into four types: nine of them are shopping mall parking lots, five are in working and business area, and two of them are located in the residential area. Then, three of the parking lots are in the comprehensive service areas, including shopping, business, working region, and even areas for transit sharing. All of the parking lots are open to the public. The formats of the original data records are vehicle parking information, including parking start time, end time, parking duration, parking price, and vehicle information. Then, the authors processed the parking records into parking lot occupancy-based data, 24/7. The records are from March 1 to Aug 31[t] of 2018. The time gap between the occupancy records is five min. In addition, the team collected real-time weather information for the same time period with a time interval of five min. Eight categories of weather conditions were summarized for future analysis: fair, cloudy, light rain, rain, heavy rain, light snow, snow, and fog.

Truck Parking Data: In the research, 49 truck parking slots data in 2 rest areas near the Interstate 5 freeway from January 01 to March 31 of 2020, and of 2021, was collected. The data are summarized into two formats, parking event data, and occupancy data. The event data is used to record each parking slot status, which includes the status change, the starting time of parking, and the time duration. For the occupancy data, the team processed the event data and obtained the occupancy rate of each parking lot based on the per minute interval. Moreover, the team also collected real-time weather information from the closest weather station during the same period and recorded every min. Eight categories of weather conditions were summarized for future analysis: cloudy, light rain, light snow, rain, heavy rain, snow, fair, and fog.

In the experiment, several findings are worth mentioning. First, the sequential-based prediction algorithms (RNN, LSTM, PewLSTM) work better for short-term feature extraction and prediction (5 and 10 min) rather than long-term prediction (1 and 2 h). Second, except PAP, the Google PDE has the best performance for long-term feature capturing and prediction results. Sometimes, the result is even better than the PAP with no ATEC. So, instead of sending the records directly into the sequential model, the Google PDE transforms the original input into multiple difficulty levels. This approach could balance the temporal features and general parking pattern features that make the network capture more long-term patterns for forecasting. However, such a method also reduces some of the accuracy for short-term prediction. Third, the ATEC shows a significant impact on the performance improvement of both datasets. With ATEC, attributes information is sufficiently integrated into the short-term and long-term parking prediction tasks. As Fig. 8.5 shows, even the weekday and weekend patterns are very different in some cases, and the PAP can also capture the pattern for both 10 min and 1 h. Then, the residential area always is the most predictable parking lot for every method.

8.8   Case study: transfer learning for nighttime traffic detection

Transfer learning can be adopted for dealing with challenging traffic scenarios such as adverse weather and occlusion. Challenging scenarios are rare compared to normal cases and often with a small number of labeled datasets, which further increases the challenge in traffic operation tasks. A typical example is vehicle detection at nighttime. The benchmark vehicle detection datasets are based mostly on detection in daytime (Luo et al., 2018); the nighttime images also pose greater challenges in vehicle detection with low lighting conditions and degraded feature representations for learning.

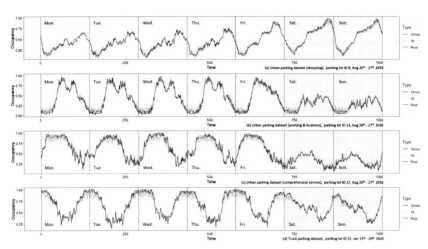

FIGURE 8.5 The PAP prediction result visualization for parking lot 08, 13, 17, 21.

Li et al. (2021a) developed a transfer learning based pipeline for vehicle detection in daytime and nighttime, using only labeled daytime images. Faster R-CNN was chosen as their vehicle detector, and they applied domain adaptation (DA) with Faster R-CNN for nighttime vehicle detection. Style transfer technique was used in the DA process, transferring the knowledge of labeled daytime images (source domain) to unlabeled nighttime images (target domain). This work also showcased that DA can be used for traffic flow parameter estimation at nighttime with a framework that compared daytime and nighttime traffic flow at the same location.

Faster R-CNN has been a widely used object detection since 2015 (Ren et al., 2015). It is a two-stage algorithm with stage one *region proposal network* and stage two *classification*. Faster R-CNN detectors were trained using daytime images to conduct two vehicle-detection tasks, one for daytime vehicle detection and another for nighttime detection. Daytime detection is a standard supervised learning task for training and testing both in the same domain. The second task is about training a Faster R-CNN model for nighttime vehicle detection without prelabeled data in nighttime training images. Style transfer was adopted as the first component to translate the daytime images to synthetic nighttime images with the consideration of daytime and nighttime image styles. The style transfer method basically learned the unpaired image-to-image translation between two different domains using CycleGAN proposed in 2017 (Zhu et al., 2017).

In Fig. 8.6, the domain for generated synthetic images is highlighted with a hat. $\widehat{T}$ is the domain for generated synthetic nighttime images from actual daytime images and $\widehat{S}$ the domain for generated synthetic daytime images from actual nighttime images.

Ideally, for one image $i_S \in I_s$, it can be translated to a synthetic image in $\widehat{T}$ by the generator $G_{S \to T}$. The adversarial discriminator D_T will encourage

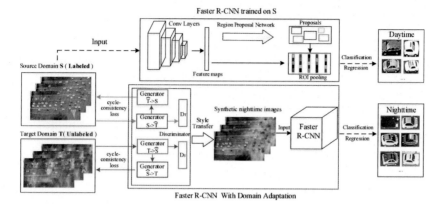

FIGURE 8.6 Faster R-CNN with DA (from Li et al., 2021a).

the translated image indistinguishable from the domain **T**. After translating the synthetic image back to the domain **S** by $G_{T \to S}$, it produced a reconstructed image $G_{T \to S}(G_{S \to T}(i_S))$ that should be similar to the original image i_S. In other words, the reconstruction error for i_S should be minimized when training the GAN, so is that for the image i_T. This reconstruction error is called cycle consistency loss, and this algorithm can be applied to unpaired image-to-image style transfer. The total loss function in the style transfer architecture is denoted as

$$
L_{\text{CycleGAN}}(G_{S \to T}, G_{T \to S}, D_S, D_T, S, T) = L_{GAN}(G_{S \to T}, D_T, S, T) + \\
L_{GAN}(G_{T \to S}, D_S, T, S) + \lambda L_{\text{Cycle}}(G_{S \to T}, G_{T \to S}, S, T),
$$

where λ is the balance weight, L_{Cycle} is the cycle consistency loss in the cycle architecture, and L_{GAN} is the adversarial training loss. The cycle consistency loss is used to regularize the GAN training. The cycle consistent loss is an L_1 penalty in the cycle architecture, which is defined as

$$
L_{\text{Cyde}}(G_{S \to T}, G_{T \to S}, S, T) = E_{i_S \sim I_S}\left[\|G_{T \to S}(G_{S \to T}(i_S)) - i_S\|_1\right] \\
+ E_{i_T \sim I_T}\left[\|G_{S \to T}(G_{T \to S}(i_T)) - i_T\|_1\right].
$$

The adversarial training loss is defined as:

$$
L_{GAN}(G_{S \to T}, D_T, S, T) = E_{i_T \sim I_T}\left[\log(D_T(i_T))\right] + \\
E_{i_S \sim I_S}\left[\log(1 - D_T(G_{S \to T}(i_S))\right].
$$

The synthetic nighttime images generated from daytime images are very similar to actual nighttime images, sharing the same vehicle locations so that the labeled vehicle bounding boxes in the source domain **S** work for the synthetic nighttime images as the training set for the Faster R-CNN.

8.9 Case study: simulation to real-world knowledge transfer for driving behavior recognition

Collecting sufficient data in the real world is sometimes challenging, due either to the raw data collection itself (e.g.,crash data) or the lack of data labels (e.g., labeling data in dash cam). Generating data from simulation software is an alternative. Driving simulators are used in many studies as surrogate data sources for different transportation applications. Macroscopic and microscopic traffic simulators (e.g., SUMO, VISSIM) are also commonly used in generating various types of traffic data for real-world scenario analysis and modeling. However, the machine learning model trained in a simulation environment cannot generalize well to the real-world problem due to domain gaps. A knowledge transfer framework is often required.

In this case study, we selected the paper "Virtual-to-real knowledge transfer for driving behavior recognition: framework and a case study" (Lu et al., 2019) as a representative example to introduce knowledge transfer from simulation to physical world. This paper uses driving simulator data for driver behavior recognition, specifically, lane changing- behavior classification. Two transfer learning methods, i.e., semi-supervised manifold alignment (SMA) and kernel manifold alignment (KEMA) are adopted to map simulation data to the real world.

The framework proposed by Lu et al. is shown in Fig. 8.7. The behavior recognition problem can be defined as a pattern classification problem. In this case, there are two classes, i.e., lane keeping and lane changing. The classifier trained using driving simulator data is expected to do the classification task in the real world. The key step in virtual-to-real knowledge transfer is to align the source domain and target domain to the same latent space. In other words, the simulation data with sufficient labels and the real data with few labels will be transferred to the same latent space using proposed techniques, e.g., manifold alignment.

In the lane-changing scenario considered in this paper, two kinds of labels, namely lane keeping (LK) and lane changing (LC) are given to distinguish the different behaviors of drivers. The data collected from the simulator and real driving scene have the same structure and can be presented by the following column vector:

$$\mathbf{d}_t = [\underbrace{x_{h,t}\, y_{h,t}\, v_{h,t}\, \theta_{h,t}}_{\text{host vehicle}}\ \underbrace{\mathbf{x}_{o,t}\, \mathbf{y}_{o,t}\, \mathbf{v}_{o,t}\, \boldsymbol{\theta}_{o,t}}_{\text{other vehicles}}]^T,$$

where $\mathbf{d}_t$ is the data instance collected at time step t, $x_{h,t}$ and $y_{h,t}$ is the longitudinal and lateral position of the host vehicle, respectively; $\theta_{h,t}$ is the heading angle of the host vehicle; and $v_{h,t}$ is the speed of the host vehicle. $\mathbf{x}_{o,t}$, $\mathbf{y}_{o,t}$, $\mathbf{v}_{o,t}$ and $\boldsymbol{\theta}_{o,t}$ are row vectors containing the longitudinal position, lateral position, speed, and heading information for other vehicles.

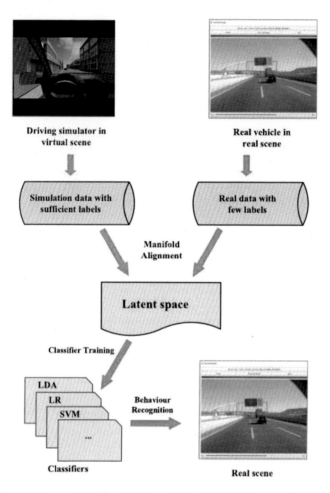

FIGURE 8.7 The framework for knowledge transfer (Lu et al., 2019).

Two datasets $\mathbf{D}^s = \left\{\mathbf{d}_i^s\right\}_{i=1}^{N_s}$ and $\mathbf{D}^r = \left\{\mathbf{d}_j^r\right\}_{j=1}^{N_r}$ containing the data collected from simulator $\left(\mathbf{d}_i^s\right)$ and real-driving scene $\left(\mathbf{d}_j^r\right)$ can be obtained. Abundant data in $\mathbf{D}^s$ are assigned with data labels *lane keeping* or *lane changing*, yet only few data in $\mathbf{D}^r$ have the label information. The target of the transfer learning framework is to train a good classifier using $\mathbf{D}^s$ and $\mathbf{D}^r$, and adopt this classifier for lane-changing behavior recognition in $\mathbf{D}^r$. The combined dataset D with the dimensionality $\left(N_{d,s} + N_{d,r}\right) \times \left(N_s + N_r\right)$ can be produced with:

$$\mathbf{D} = \begin{bmatrix} \mathbf{D}^s & \mathbf{0} \\ \mathbf{0} & \mathbf{D}^r \end{bmatrix}.$$

The semi-supervised manifold alignment (SMA) was proposed in Wang and Mahadevan (2011), which can reuse data (either labeled or not) collected from the source domain to improve classification performances in the target domain. As aforementioned, manifold alignment can map multiple domains into one latent space. In this case study, the latent manifold $\mathbf{M}$ has the form:

$$\mathbf{M} = \left[\mathbf{M}^s, \mathbf{M}^r\right],$$

where $\mathbf{M}^s$ and $\mathbf{M}^r$ contain the data mapped from the original simulation and real domain respectively. Two mapping matrices $\mathbf{Q}^s$ and $\mathbf{Q}^r$ can be used to map the data from original source domains to $\mathbf{M}^s$ and $\mathbf{M}^r$. Integrating $\mathbf{Q}^s$ and $\mathbf{Q}^r$, the following joint mapping matrix as follows is acquired:

$$\mathbf{Q} = \left[\begin{array}{c} \mathbf{Q}^s \\ \mathbf{Q}^r \end{array}\right].$$

The goal of SMA is to find an optimal mapping matrix $\mathbf{Q}^*$ that minimizes the following cost function:

$$J(\mathbf{Q}) = \frac{\mu_J J_{\text{geo}}(\mathbf{Q}) + J_{\text{sim}}(\mathbf{Q})}{J_{dis}(\mathbf{Q})}.$$

The detailed mathematical derivation of the solution process can be found in the original paper. After obtaining the optimal mapping matrix Q^*, the data from the simulation and real domains can be mapped to the same latent manifold by:

$$\mathbf{M} = \left[\mathbf{M}^s, \mathbf{M}^r\right] = \left(\mathbf{Q}^*\right)^T \mathbf{D}.$$

8.10 Exercises

- What is the advantage of transfer learning over traditional supervised learning?
- Explain inductive transfer learning, transductive transfer learning, and unsupervised transfer learning.
- Explain self-taught learning, multitask learning, domain adaptation, and sample selection bias/covariance shift.
- What is the difference between few-shot learning and zero-shot learning? Give an example of traffic video analytics.
- Two techniques in transfer learning, freezing and fine-tuning, have different impacts. In what situations should you choose freezing over fine-tuning?
- Why is labeled data from traffic surveillance images critical for the performance of traffic video analytics–since there are already a large number of labeled vehicle datasets?

- If you are a transfer learning engineer and asked to design a detector that takes images from simulation for real-world traffic video detection, how would you design the transfer learning method?
- What is the key knowledge needed for domain adaptation from daytime images to nighttime images?
- Explain the usage of latent space in simulation to real-world domain adaption.

Chapter 9

Graph neural networks

Graph neural networks (GNNs) are neural network models that capture the dependence of graphs via messages passing between the nodes of graphs. In recent years, variants of GNNs, such as graph convolutional network (GCN) and graph attention network (GAT), have demonstrated ground-breaking performances on many deep learning tasks, including traffic road-network modeling. In this chapter, we will introduce the basics of graph neural networks and several representative graph neural networks. Graph-related case studies will be introduced to show how GNNs can be applied to transportation tasks.

9.1 Preliminaries

In this section, we introduce models of GNNs in a designer view. We first present the general design pipeline for designing a GNN model in this section.

In later sections, we denote a graph as $G = (V, E)$, where $|V| = N$ is the number of nodes in the graph and $|E| = N^e$ is the number of edges. $\mathbf{A} \in \mathbb{R}^{N \times N}$ is the adjacency matrix. For graph representation learning, we use $\mathbf{h}_v$ and $\mathbf{o}_v$ as the hidden state and output vector of node v.

9.2 Graph neural networks

The convolution operators that we introduce in this section are the mostly used propagation operators for GNN models. The main idea of convolution operators is to generalize convolutions from other domain to the graph domain. Advances in this direction are often categorized as either spectral approaches or spatial approaches.

9.2.1 Spectral GNN

Spectral approaches work with a spectral representation of the graphs. These methods are theoretically based on graph signal processing (Shuman et al., 2013) and define the convolution operator in the spectral domain.

In spectral methods, a graph signal $\mathbf{x}$ is first transformed to the spectral domain by the graph Fourier transform $\mathcal{F}$, and then the convolution operation is conducted. After the convolution, the resulted signal is transformed back using the inverse graph Fourier transform $\mathcal{F}^{-1}$. These transforms are defined as

$$\mathcal{F}(\mathbf{x}) = \mathbf{U}^T \mathbf{x} \tag{9.1}$$

Machine Learning for Transportation Research and Applications
https://doi.org/10.1016/B978-0-32-396126-4.00014-X

$$\mathcal{F}^{-1}(\mathbf{x}) = \mathbf{U}\mathbf{x}. \qquad (9.2)$$

Here, $\mathbf{U}$ is the matrix of eigenvectors of the normalized graph Laplacian $\mathbf{L} = \mathbf{I}_N - \mathbf{D}^{-\frac{1}{2}}\mathbf{A}\mathbf{D}^{-\frac{1}{2}}$ ($\mathbf{D}$ is the degree matrix and, $\mathbf{A}$ is the adjacency matrix of the graph). The normalized graph Laplacian is real symmetric positive semidefinite, so it can be factorized as $\mathbf{L} = \mathbf{U}\mathbf{\Lambda}\mathbf{U}^T$ (where $\mathbf{\Lambda}$ is a diagonal matrix of the eigenvalues). Based on the convolution theorem (Mallat, 1999), the convolution operation is defined as

$$\begin{aligned}
\mathbf{g} \star \mathbf{x} &= \mathcal{F}^{-1}(\mathcal{F}(\mathbf{g}) \odot \mathcal{F}(\mathbf{x})) \\
&= \mathbf{U}\left(\mathbf{U}^T\mathbf{g} \odot \mathbf{U}^T\mathbf{x}\right),
\end{aligned} \qquad (9.3)$$

where $\mathbf{U}^T\mathbf{g}$ is the filter in the spectral domain. If we simplify the filter by using a learnable diagonal matrix $\mathbf{g}_w$, then we have the basic function of the spectral methods:

$$\mathbf{g}_w \star \mathbf{x} = \mathbf{U}\mathbf{g}_w\mathbf{U}^T\mathbf{x}. \qquad (9.4)$$

Next, we will introduce several typical spectral methods which design various filters $\mathbf{g}_w$.

Spectral Network: Spectral network (Bruna et al., 2014) uses a learnable diagonal matrix as the filter, that is $\mathbf{g}_w = \mathrm{diag}(\mathbf{w})$, where $\mathbf{w} \in \mathbb{R}^N$ is the parameter. However, this operation is computationally inefficient and the filter is nonspatially localized. Henaff et al. (2015) attempt to make the spectral filters spatially localized by introducing a parameterization with smooth coefficients.

ChebNet: Hammond et al. (2011) suggest that $\mathbf{g}_w$ can be approximated by a truncated expansion in terms of Chebyshev polynomials $\mathbf{T}_k(x)$ up to the Kth order. Defferrard et al. (2016) propose the ChebNet based on this theory. The operation can be described as

$$\mathbf{g}_w \star \mathbf{x} \approx \sum_{k=0}^{K} w_k\mathbf{T}_k(\tilde{\mathbf{L}})\mathbf{x}, \qquad (9.5)$$

where $\tilde{\mathbf{L}} = \frac{2}{\lambda_{\max}}\mathbf{L} - \mathbf{I}_N$, $\lambda_{\max}$ represents the largest eigenvalue of $\mathbf{L}$. The range of the eigenvalues in $\tilde{\mathbf{L}}$ is $[-1, 1]$. $\mathbf{w} \in \mathbb{R}^K$ is now a vector of Chebyshev coefficients. The Chebyshev polynomials are defined as $\mathbf{T}_k(\mathbf{x}) = 2\mathbf{x}\mathbf{T}_{k-1}(\mathbf{x}) - \mathbf{T}_{k-2}(\mathbf{x})$, with $\mathbf{T}_0(\mathbf{x}) = 1$ and $\mathbf{T}_1(\mathbf{x}) = \mathbf{x}$. It can be observed that the operation is K-localized since it is a Kth-order polynomial in the Laplacian. Defferrard et al. (2016) use this K-localized convolution to define a convolutional neural network that could remove the need to compute the eigenvectors of the Laplacian.

GCN: Kipf and Welling (2017) propose graph convolution network (GCN) to simplify the convolution operation in Eq. (9.5) with $K = 1$. They assume $\lambda_{\max} \approx 2$ and simplify the equation to

$$\mathbf{g}_w \star \mathbf{x} \approx w_0\mathbf{x} + w_1\left(\mathbf{L} - \mathbf{I}_N\right)\mathbf{x} = w_0\mathbf{x} - w_1\mathbf{D}^{-\frac{1}{2}}\mathbf{A}\mathbf{D}^{-\frac{1}{2}}\mathbf{x}, \qquad (9.6)$$

with two free parameters w_0 and w_1. With parameter constraint $w = w_0 = -w_1$, we can obtain the following expression:

$$\mathbf{g}_w \star \mathbf{x} \approx w \left(\mathbf{I}_N + \mathbf{D}^{-\frac{1}{2}} \mathbf{A} \mathbf{D}^{-\frac{1}{2}} \right) \mathbf{x}. \tag{9.7}$$

GCN further introduces a renormalization process in Eq. (9.7) which representing $\mathbf{I}_N + \mathbf{D}^{-\frac{1}{2}} \mathbf{A} \mathbf{D}^{-\frac{1}{2}}$ as $\tilde{\mathbf{D}}^{-\frac{1}{2}} \tilde{\mathbf{A}} \tilde{\mathbf{D}}^{-\frac{1}{2}}$, with $\tilde{\mathbf{A}} = \mathbf{A} + \mathbf{I}_N$ and $\tilde{\mathbf{D}}_{ii} = \sum_j \tilde{\mathbf{A}}_{ij}$. Finally, we can get the compact form of GCN as

$$\mathbf{H} = \tilde{\mathbf{D}}^{-\frac{1}{2}} \tilde{\mathbf{A}} \tilde{\mathbf{D}}^{-\frac{1}{2}} \mathbf{X} \mathbf{W}, \tag{9.8}$$

where $\mathbf{X} \in \mathbb{R}^{N \times F}$ is the input matrix, $\mathbf{W} \in \mathbb{R}^{F \times F'}$ is the parameter and $\mathbf{H} \in \mathbb{R}^{N \times F'}$ is the convolved matrix. F and F' are the dimensions of the input and the output, respectively. Note, you can find a lot of GCN variants that are proposed to learn comprehensive features in the graph structure flexibly and efficiently.

9.2.2 Spatial GNN

Spatial approaches define convolutions directly on the graph based on the graph topology. The idea of spatial GNN operation is similar to that of the CNN operation that targets to focus on the localized features of the input data as shown in Fig. 9.1. The major challenge of CNNs is to define the convolution operation with differently sized neighborhoods and maintain the local invariance. Specifically, spatial GNN encounters the challenges.

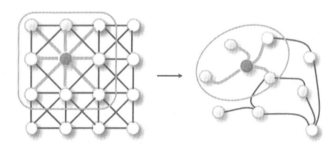

FIGURE 9.1 Similarity of the ideas of CNN and GNN.

DCNN. The diffusion convolutional neural network (DCNN) (Atwood and Towsley, 2016) uses transition matrices to define the neighborhood for nodes. The diffusion representations of each node in the graph can be expressed as

$$\mathbf{H} = f (\mathbf{W}_c \odot \mathbf{P} \mathbf{X}) \in \mathbb{R}^{N \times K \times F}, \tag{9.9}$$

where the $\odot$ operator represents element-wise multiplication, $\mathbf{X} \in \mathbb{R}^{N \times F}$ is the input matrix, and F is the dimension of features. $\mathbf{P}^*$ is an $N \times K \times N$ tensor, which contains the power series $\{\mathbf{P}, \mathbf{P}^2, \ldots, \mathbf{P}^K\}$ of matrix $\mathbf{P}$. And $\mathbf{P}$ is the degree-normalized transition matrix derived from the graphs adjacency matrix **A**. Each entity is transformed to a diffusion convolutional representation, which is a $K \times F$ matrix defined by K hops of graph diffusion over F features. It will be defined by a $K \times F$ weight matrix W_c followed by the nonlinear activation function f.

9.2.3 Attention-based GNNs

The attention mechanism has been successfully used in many transportation-related tasks, such as traffic forecasting (Park et al., 2020), travel time estimation (Fang et al., 2020), autonomous driving (Cai et al., 2022), and so on. There are also several models that try to generalize the attention operator on graphs (Veličković et al., 2018a). Compared with the operators introduced before, attention-based operators assign different weights for neighbors, so that they can alleviate noises and achieve better results.

GAT. The graph attention network (GAT) (Veličković et al., 2018b) incorporates the attention mechanism into the propagation step. It computes the hidden states of each node by attending to its neighbors, following a self-attention strategy. The hidden state of node v can be obtained by

$$\mathbf{h}_v^{t+1} = \rho \left(\sum_{u \in r_v} \alpha_{vu} \mathbf{W} \mathbf{h}_u^t \right) \tag{9.10}$$

$$\alpha_{vu} = \frac{\exp\left(\text{LeakyReLU}\left(\mathbf{a}^T \left[\mathbf{W}\mathbf{h}_v \| \mathbf{W}\mathbf{h}_u\right]\right)\right)}{\sum_{k \in r_v} \exp\left(\text{LeakyReLU}\left(\mathbf{a}^T \left[\mathbf{W}\mathbf{h}_v \| \mathbf{W}\mathbf{h}_k\right]\right)\right)}, \tag{9.11}$$

where $\mathbf{W}$ is the weight matrix associated with the linear transformation that is applied to each node and a is the weight vector of a single-layer MLP.

Moreover, GAT utilizes the multi-head attention used by (Vaswani et al., 2017) to stabilize the learning process. It applies K independent attention-head matrices to compute the hidden states and then concatenates their features, resulting in the following output representations:

$$\mathbf{m}_v^{t+1} = \sum_{u \in Y_v} M_t \left(\mathbf{h}_v^t, \mathbf{h}_u^t, \mathbf{e}_{vu}\right)$$
$$\mathbf{h}_v^{t+1} = U_t \left(\mathbf{h}_v^t, \mathbf{m}_v^{t+1}\right). \tag{9.12}$$

Here, $\mathbf{e}_{vu}$ represents features of undirected edge (v, u).

$$\mathbf{h}_v^{t+1} = \|_{k=1}^K \sigma \left(\sum_{u \in V_v} \alpha_{vu}^k \mathbf{W}_k \mathbf{h}_u^t \right)$$
$$\mathbf{h}_v^{t+1} = \sigma \left(\frac{1}{K} \sum_{k=1}^K \sum_{u \in I_v} \alpha_{vu}^k \mathbf{W}_k \mathbf{h}_u^t \right). \tag{9.13}$$

Here ∥ is the concatenate operation, and α_{uv}^k is the normalized attention coefficient computed by the k-th attention mechanism a^k. The attention architecture has several properties: (1) The attention mechanism can work in a parallel way that makes the computation of the graph attention network highly efficient; (2) applying attention to any setting makes the capacity and accuracy of the model very high because the models need to learn only important data (less amount of data); and (3) the learned weights in GAT can make the process of the network more interpretable.

9.3 Case study 1: traffic graph convolutional network for traffic prediction

9.3.1 Problem definition

Traffic forecasting refers to predicting future traffic states, such as traffic speed, travel time, or volume, given previously observed traffic states from a road network, as introduced in the RNN chapter. The traffic network can be converted into a topological graph consisting of all N nodes, representing N traffic sensing locations, and a set of edges. The short-term traffic forecasting problem aims to learn a function $F(\cdot)$ to map T time steps of historical graph signals, i.e., $X_T = [x_1, \ldots, x_t, \ldots, x_T]$, to the graph signals in the subsequent one or multiple time steps. In this study, the function attempts to forecast the graph signals in the subsequent one step, i.e., x_{T+1}, and the formulation of $F(\cdot)$ is defined as

$$F\left([x_1, \ldots, x_t, \ldots, x_T]; G\left(\mathcal{V}, \mathcal{E}, \tilde{A}^k\right)\right) = x_{T+1}. \tag{9.14}$$

Notations in Roadway Graph: The traffic network and the relationship between roadway locations can be represented by an undirected graph $\mathcal{G}$, where $\mathcal{G} = (\mathcal{V}, \mathcal{E})$ with N nodes $v_i \in \mathcal{V}$ and edges $(v_i, v_j) \in \mathcal{E}$. Even though some roads are directed in the reality, due to the impact of traffic congestion occurring on these roads will be bidirectionally propagated to upstream and downstream roads, we take the bidirectional impact into account, and thus let $\mathcal{G}$ be an undirected graph. The connectedness of nodes in $\mathcal{G}$ is represented by an adjacency matrix $A \in \mathbb{R}^{N \times N}$, in which each element $A_{i,j} = 1$ if there is an edge connecting node i and node j and $A_{i,j} = 0$, otherwise $(A_{i,i} = 0)$. Based on the adjacency matrix, the degree matrix of $\mathcal{G}$, which measures the number of edges attached to each vertex, can be defined as $D \in \mathbb{R}^{N \times N}$ in which $D_{ii} = \sum_j A_{ij}.D$ is a diagonal matrix and all non-diagonal elements are zeros.

Based on the adjacency matrix, an edge counting function $d(v_i, v_j)$ can be defined as counting the minimum number of edges traversed from node i to node j. Then, the set of k-hop (k-th order) neighborhoods of each node i, including node i itself, can be defined as $\{v_j \in \mathcal{V} \mid d(v_i, v_j) \leq k\}$. However, since the traffic states are time series data and the current traffic state on a road will definitely influence the future state, we consider the all roads are self-influenced. Thus,

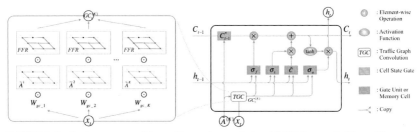

FIGURE 9.2 Basic structure of the traffic graph convolutional LSTM network. The traffic graph convolution (TGC) is shown on the left-hand side in detail by unfolding the traffic graph convolution at time t, in which $\tilde{A}^k$ s and $\mathcal{FFR}$ with respect to a red star node are demonstrated.

we consider the neighborhood of a node contains the node itself and a neighborhood matrix to characterize the one-hop neighborhood relationship of the whole graph, denoted as $\tilde{A} = A + I$, where I is the identity matrix. Then, the k-hop neighborhood relationship of the graph nodes can be characterized by $(A + I)^k$. However, some elements in $(A + I)^k$ will inevitably exceed one. Owing to the k-hop neighborhood of a node is only used for describing the existence of all the k-hop neighbors, it is not necessary to make a node's k-hop neighbors weighted by the number of hops. Thus, we clip the values of all elements in $(A + I)^k$ to be in $\{0, 1\}$ and define a new k-hop neighborhood matrix $\tilde{A}^k$, in which each element $\tilde{A}^k_{i,j}$ satisfies

$$\tilde{A}^k_{i,j} = \min\left((A + I)^k_{i,j}, 1\right),$$ (9.15)

where min refers to minimum. In this case, $\tilde{A}^1 = A^1 = A$. An intuitive example of k-hop neighborhood with respect to a node (a red star) is illustrated by the blue points (mid gray in print version) on the left-hand side of Fig. 9.2.

Free-Flow Reachable Matrix: Based on the length of each road in the traffic network, we define a distance matrix $\text{Dist} \in \mathbb{R}^{N \times N}$, where each element $\text{Dist}_{i,j}$ represents the real roadway distance from node i to j ($i \neq j$). When taking the underlying physics of vehicle traffic on a road network into consideration, we need to understand that the impact of a roadway segment on adjacent segments is transmitted in two primary ways: 1) slowdowns and/or blockages propagating upstream; and 2) driver behavior and vehicle characteristics associated with a particular group of vehicles traveling downstream. Thus, for a traffic network-based graph or other similar graphs, the traffic impact transmission between nonadjacent nodes cannot bypass the intermediate node/nodes, and, thus, we need to consider the reachability of the impact between adjacent and nearby node pairs. To ensure the traffic impact transmission between k-hop adjacent nodes follow the established traffic flow theory (Daganzo, 1994), we define a free-flow reachable matrix, $\mathcal{FFR} \in \mathbb{R}^{N \times N}$, that

$$\mathcal{FFR}_{i,j} = \begin{cases} 1, & S_{i,j}^{\mathcal{FF}} m \Delta t - \text{Dist}_{i,j} \geq 0 \\ 0, & \text{otherwise} \end{cases}, \quad \forall v_i, v_j \in \mathcal{V}, \tag{9.16}$$

where $S_{i,j}^{\mathcal{FF}}$ is the free-flow speed between node i and j and free-flow speed refers to the average speed that a motorist would travel if there were no congestion or other adverse conditions (such as severe weather). Δt is the duration of time quantum, and m is a number counting how many time intervals are considered to calculate the distance traveled under free-flow speed. Thus, m determines the temporal influence of formulating the $\mathcal{FFR}$. Each element $\mathcal{FFR}_{i,j}$ equals one if vehicles can traverse from node i to j in m time-step, $m \cdot \Delta t$, with free-flow speed, and $\mathcal{FFR}_{i,j} = 0$, otherwise. Intuitively, the $\mathcal{FFR}_{i,j}$ measures whether a vehicle can travel from node i to node j with the free-flow speed under a specific time interval. We consider each road is self-reachable, and thus, all diagonal values of $\mathcal{FFR}$ are set as one. An example, $\mathcal{FFR}$ with respect to a node (a red star), is shown by green lines (dark gray in print version) on the left side of Fig. 9.2.

Traffic Graph Convolution: As we introduced in the Spatial GNN section, GNN can extend the receptive field of graph convolution by replacing the one-hop neighborhood matrix $\tilde{A}$ with the k-hop neighborhood matrix $\tilde{A}^k$. However, previous studies either neglect the properties of the edges in a graph, such as the distances between various sensing locations (the lengths of the graph edges) and the free-flow reachability or fail to consider high-order neighborhood of nodes in the graph. Hence, to comprehensively solve the network-wide forecasting problem, we consider both graph edge properties and high-order neighborhood in the traffic network-based graph. Hence, we define the k-order (k-hop) Traffic Graph Convolution (TGC) operation as

$$GC_t^k = \left(W_{gc_k} \odot \tilde{A}^k \odot \mathcal{FFR} \right) x_t, \tag{9.17}$$

where $\odot$ is the Hadamard product operator, i.e., the element-wise matrix multiplication operator, and $x_t \in \mathbb{R}^N$ is the vector of traffic states (speed) of all nodes at time t. The $W_{gc\_k} \in \mathbb{R}^{N \times N}$ is a trainable weight matrix for the k-order traffic graph convolution and the $GC^k \in \mathbb{R}^N$ is the extracted k-order traffic graph convolution feature. Due to the fact that $\tilde{A}^k$ and $\mathcal{FFR}$ are both sparse matrices containing only the elements 0 and 1, the result of $W_{gc_k} \odot \tilde{A}^k \odot \mathcal{FFR}$ is also sparse. Further, the trained weight W_{gc_k} has the potential to measure the interactive influence between graph nodes and, thus, enhance the interpretability of the model.

Let $K \leq K_{\max}$ denote the largest hop for traffic graph convolution in this study, and the corresponding traffic graph-convolution feature is GC_t^K with respect to input data x_t. Different hops of neighborhood in TGC will result in different extracted features. To enrich the feature space, the features extracted from different orders (from 1 to K) of traffic graph convolution with respect to

X_t are concatenated together as a vector defined as follows

$$\boldsymbol{GC}_t^{\{K\}} = \left[GC_t^1, GC_t^2, \ldots, GC_t^K \right]. \tag{9.18}$$

The $\boldsymbol{GC}_t^{\{K\}} \in \mathbb{R}^{N \times K}$ contains all the K orders of traffic graph convolutional features, as intuitively shown on the left-hand side of Fig. 9.2. In this study, after operating the TGC on input data x_t, the generated $\boldsymbol{GC}_t^{\{K\}}$ will be fed into the following layer in the proposed neural network structure described in the following section.

9.3.2 Method: traffic graph convolutional LSTM

We propose a Traffic Graph Convolutional LSTM (TGCLSTM) recurrent neural network, as shown on the right-hand side of Fig. 9.2, which learns both the complex spatial dependencies and the dynamic temporal dependencies presented in traffic data. In this model, the gates structure in the vanilla LSTM and the hidden state are unchanged, but the input is replaced by the graph convolution features, which are reshaped into a vector $\boldsymbol{GC}^{\{K\}} \in \mathbb{R}^{KN}$. The forget gate f_t, the input gate i_t, the output gate o_t, and the input cell state $\tilde{C}_t$ in terms of time step t are defined as follows:

$$
\begin{aligned}
\mathrm{f}_t &= \sigma_g \left(W_f \cdot \boldsymbol{GC}_t^{\{K\}} + U_f \cdot h_{t-1} + b_f \right) \\
\mathrm{i}_t &= \sigma_g \left(W_i \cdot \boldsymbol{GC}_t^{\{K\}} + U_i \cdot h_{t-1} + b_i \right) \\
\mathrm{o}_t &= \sigma_g \left(W_o \cdot \boldsymbol{GC}_t^{\{K\}} + U_o \cdot h_{t-1} + b_o \right) \\
\tilde{C}_t &= \tanh \left(W_C \cdot \boldsymbol{GC}_t^{\{K\}} + U_C \cdot h_{t-1} + b_C \right),
\end{aligned}
\tag{9.19}
$$

where $\cdot$ is the matrix multiplication operator. W_f, W_i, W_o, and $W_C \in \mathbb{R}^{N \times KN}$ are the weight matrices, mapping the input to the three gates and the input cell state, while U_f, U_i, U_o, and $U_C \in \mathbb{R}^{N \times N}$ are the weight matrices for the preceding hidden state. b_f, b_i, b_o, and $b_C \in \mathbb{R}^N$ are four bias vectors. The σ_g is the gate activation function, which typically is the sigmoid function, and tanh is the hyperbolic tangent function. A cell state gate is designed and added in the LSTM cell. The cell state gate is defined as follows

$$C_{t-1}^* = W_{\mathcal{N}} \odot \left(\tilde{A}^K \odot \mathcal{FFR} \right) \cdot C_{t-1}, \tag{9.20}$$

where $W_{\mathcal{N}}$ is a weight matrix to measure the contributions of neighboring cell states. To correctly reflect the traffic network structure, the $W_{\mathcal{N}}$ is constrained by multiplying a $\mathcal{FFR}$ based K-hop adjacency matrix, $\tilde{A}^K \odot \mathcal{FFR}$. With this gate, the influence of neighboring cell states will be considered when the cell state is recurrently input to the subsequent time step. Then, the final cell state and the hidden state are calculated as follows

$$C_t = f_t \odot C_{t-1}^* + i_t \odot \tilde{C}_t$$
$$h_t = o_t \odot \tanh(C_t).$$

(9.21)

At the final time step T, the hidden state h_T is the output of TGC-LSTM, namely, the predicted value $\hat{y}_T = h_T$. Let $y_T \in \mathbb{R}^N$ denote the label of the input data $X_T \in \mathbb{R}^{N \times N}$. For the sequence prediction problem in this study, the label of time step T is the input of the next time step $(T+1)$ such that $y_T = x_{T+1}$. Then, the loss during the training process is defined as

$$\text{Loss} = L\left(y_T, \hat{y}_T\right) = L\left(x_{T+1}, h_T\right),$$

(9.22)

where $L(\cdot)$ is a function to calculate the residual between the predicted value $\hat{y}_T$ and the true value y_T. Normally, the $L(\cdot)$ function is a Mean Squared Error (MSE) function for predicting continuous values. Since the proposed model contains a traffic graph-convolution operation, the generated set of TGC features $GC_t^{\{K\}}$ and the learned TGC weights $\left\{W_{gc_1}, \ldots, W_{gc_K}\right\}$ provide an opportunity to make the proposed model interpretable via analyzing the learned TGC weights.

9.3.3 Results

Dataset: In this case study, two real-world network-scale traffic-speed datasets are utilized. The first contains data collected from inductive loop detectors deployed on four connected freeways in the Greater Seattle area, shown in Fig. 9.3(a). This dataset contains traffic state data from 323 sensor stations over the entirety of 2015 at 5-minute intervals. The second contains road link-level traffic speeds aggregated from GPS probe data collected by commercial vehicle fleets and mobile apps provided by the company INRIX. The INRIX traffic network covers the Seattle downtown area, shown in Fig. 9.3(b). This dataset describes the traffic state at 5-minute intervals for 1,014 road segments and covers the entire year of 2012.

Prediction Results and Visualization: Fig. 9.4 visualizes the predicted traffic speed sequences and the ground truth of two locations selected from the LOOP and INRIX dataset. Though the traffic networks of the two datasets are very different, the curves demonstrate that the trends of the traffic speed are predicted well at both peak traffic and off-peak hours.

Model Interpretation: To better understand the contribution of the graph convolution weight, the TGCLSTM model's TGC weight in Fig. 9.5(a) and its corresponding physical locations on the actual map in Fig. 9.5(b), by highlighting them with Roman numerals and red boxes. The influence of these marked weights on neighboring nodes in the LOOP data are visualized by circles. The darkness of the green and pink colors and the sizes of the circles represent the magnitude of influence. We can find in Fig. 9.5(b) that the area tagged with (IV)

FIGURE 9.3 Two datasets for testing the TGCLSTM.

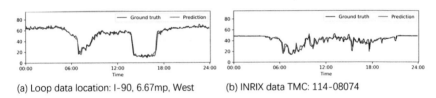

(a) Loop data location: I-90, 6.67mp, West   (b) INRIX data TMC: 114-08074

FIGURE 9.4 Prediction results of TGCLSTM at two sites from the two dataset.

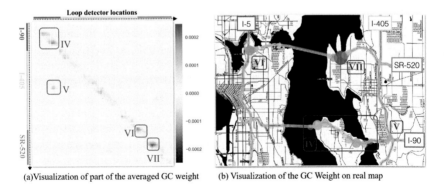

(a)Visualization of part of the averaged GC weight   (b) Visualization of the GC Weight on real map

FIGURE 9.5 Visualization of TGC weight matrix.

is quite representative because the two groups of circles are located at the inter-sections between freeways and two main corridors that represent the entrances to an island (Mercer Island). Areas (V) and (VI) are the intersections between I-90 and I-405 and between I-5 and SR-520, respectively. The VII area located on

SR-520 contains a frequent-congested ramp connecting to the city of Bellevue, the location of which is highlighted by the largest green circle.

9.4 Case study 2: graph neural network for traffic forecasting with missing values

9.4.1 Problem definition

A traffic network normally consists of multiple roadway links. The traffic forecasting task is to predict the future traffic states of all (road) links or sensor stations in the traffic network based on historical traffic state data. The collected spatiotemporal traffic state data of a traffic network with S links/sensor-stations can be characterized as a T-step sequence $[x_1, x_2, ..., x_t, ..., x_T] \in \mathbb{R}^{T \times S}$, in which $x_t \in \mathbb{R}^S$ demonstrates the traffic states of all S links at the t-th step. The traffic state of the s-th link at time t is represented by x_t^s.

Since the traffic network is composed of road links and intersections, it is intuitive to consider the traffic network as an undirected graph consisting of vertices and edges. The graph can be denoted as $\mathcal{G} = (\mathcal{V}, \mathcal{E}, \mathcal{A}, \mathcal{D})$ with a set of vertices $\mathcal{V} = \{v_1, ..., v_S\}$ and a set of edges $\mathcal{E}$ between vertices. $\mathcal{A} \in \mathbb{R}^{S \times S}$ is a symmetric (typically sparse) adjacency matrix with binary elements, where $\mathcal{A}_{i,j}$ denotes the connectedness between nodes v_i and v_j. The existence of an edge is represented through $\mathcal{A}_{i,j} = \mathcal{A}_{j,i} = 1$, otherwise $\mathcal{A}_{i,j} = 0$ ($\mathcal{A}_{i,i} = 0$). Based on $\mathcal{A}$, a diagonal graph degree matrix $\mathcal{D} \in \mathbb{R}^{S \times S}$ describing the number of edges attached to each vertex can be obtained by $\mathcal{D}_{i,i} = \sum_{j=1}^{S} \mathcal{A}_{i,j}$.

Traffic state data can be collected by multi-types of traffic sensors or probe vehicles. When traffic sensors fail or no probe vehicles go through road links, the collected traffic state data may have missing values. We use a sequence of masking vectors $[m_1, m_2, ..., m_T] \in \mathbb{R}^{T \times S}$, where $m_t \in \mathbb{R}^S$, to indicate the position of the missing values in traffic state sequence $[x_1, x_2, ..., x_T]$. The masking vector can be obtained by

$$m_t^s = \begin{cases} 1, & \text{if } x_t^s \text{ is observed} \\ 0, & \text{otherwise,} \end{cases} \tag{9.23}$$

where x_t^s is the traffic state of s-th link at step t.

Taking the missing values into consideration, we can formulate the traffic forecasting as follows:

$$F(\mathcal{G}, [x_1, x_2 ..., x_T], [m_1, m_2 ..., m_T]) = [x_{T+1}]. \tag{9.24}$$

A traffic network is a dynamic system, and the states on all links keep varying resulting from the movements of vehicles in the system. Thus, we assume the traffic network's dynamic process satisfies the Markov property that the future state of the traffic network is conditional on the present state.

Markov property: The future state of the traffic network x_{t+1} depends only upon the present state x_t, not on the sequence of states that preceded it.

Taking $X_1, X_2, ..., X_{t+1}$ as random variables with the Markov property and $x_1, x_2, ..., x_{t+1}$ as the observed traffic states, the Markov process can be formulated in a conditional probability form as

$$Pr(X_{t+1} = x_{t+1} | X_1 = x_1, X_2 = x_2, ..., X_t = x_t) = Pr(X_{t+1} = x_{t+1} | X_t = x_t), \tag{9.25}$$

where $Pr(\cdot)$ demonstrates the probability.

However, the transition matrix is temporal dependent since, at different times of the day, the traffic state's transition pattern should be different. Based on Eq. (9.25), the transition process of traffic states can be formulated in the vector form as

$$x_{t+1} = P_t x_t, \tag{9.26}$$

where $P_t \in \mathbb{R}^{S \times S}$ is the transition matrix and $(P_t)_{i,j}$ measures how much influence x_t^j has on forming the state x_{t+1}^i.

The transition process defined in Eq. (9.26) does not take the time interval between x_{t+1} and x_t into consideration. We denote the time interval between two consecutive time steps of traffic states by Δt. If Δt is small enough ($\Delta t \to 0$), the traffic network's dynamic process can be measured as a continuous process, and the differences between consecutive traffic states are close to zero, i.e., $|x_{t+\Delta t} - x_t| \to 0$. However, a long time interval may result in more variations between the present and future traffic states, leading to a more complicated transition process. Since the traffic state data are normally processed into discrete data and the size of transition matrix P_t is fixed, we consider that the longer the Δt is, the lower capability of measuring the actual transition process P_t has. Thus, we multiply a decay parameter $\gamma \in (0, 1)$ in Eq. (9.26) to represent the temporal impact on transition process as

$$x_{t+1} = \gamma P_t x_t. \tag{9.27}$$

The transition matrix can measure the contributions made by all roadway links on a specific link, which assumes that the state of a roadway link is influenced by all links in the traffic network. However, since vehicles in the traffic network traverse connected road links one by one and traffic states of connected links are transmitted by those vehicles, the traffic state of a link will be affected only by its neighboring links during a short period of time.

Graph localization property: The traffic state of a specific link s in a traffic network is mostly influenced by localized links, i.e., the link s itself and its neighboring links, during a short period of time. The neighboring links refer to the links in the graph within a specific order of hops with respect to the link s. With the help of the graph's topological structure, the localization property in the graph can be measured based the adjacency matrix in two ways: (1) The self-connection adjacency matrix **A**, describing the connectedness of vertices, can inherently indicate the localization property of all vertices in the graph.

Then, the impacts of localized links can be easily measured by a weighted self-connection adjacency matrix. (2) The other way is to conduct the spectral graph convolution operation on the traffic state x_t to measure the localized impacts in the graph. The spectral graph convolution on x_t can be defined as $U \Lambda_\theta U^T x_t$ (Defferrard et al., 2016), where U is the eigenvector matrix of the Laplacian matrix $\mathcal{L}$ and Λ_θ is a learnable diagonal weight matrix.

Graph Markov Process: With the aforementioned two properties, we define the traffic state transition process as a graph Markov process (GMP). The graph Markov process can be formulated in a conditional probability form as

$$Pr(X_{t+1} = x_{t+1}^u | X_t = x_t) = Pr(X_{t+1} = x_{t+1}^u | X_t = x_t^v, v \in \mathcal{N}(u)), \quad (9.28)$$

where the superscripts u and v are the indices of graph links (road links). The $\mathcal{N}(u)$ denotes a set of one-hop neighboring links of link u and link u itself. The properties of this graph Markov process is similar to the properties of the Markov random field (Rue and Held, 2005) with temporal information. Since the influence of a road link on its neighbors is gradually spread by the vehicles traveling on this road link, a road link's one-hop neighbors are the ones directly influence it. Thus, we assume that road links are only influenced by their one-hop neighbors in the graph. Based on Eq. (9.27), we can easily incorporate the graph localization properties into the traffic states' transition process by element-wise multiplying the transition matrix P_t with the self-connection adjacency matrix **A**. Then, the GMP can be formulated in the vector form as

$$x_{t+1} = \gamma(\mathbf{A} \odot P_t)x_t, \quad (9.29)$$

where $\odot$ is the Hadamard (element-wise) product operator that $(\mathbf{A} \odot P_t)_{ij} = \mathbf{A}_{ij} \times (P_t)_{ij}$.

As we assume the traffic state transition process follows the graph Markov process, the future traffic state can be inferred by the present state. If there are missing values in the present state, we can infer the missing values from previous states. We consider x_t is the observed traffic state at time t, and a mask vector m_t can be acquired according to Eq. (9.23). We denote the completed state by $\tilde{x}_t$, in which all missing values are filled based on historical data. Hence, the completed state consists of two parts, including the observed state values and the inferred state values, as follows:

$$\tilde{x}_t = x_t \odot m_t + \tilde{x}_t \odot (1 - m_t), \quad (9.30)$$

where $\tilde{x}_t \odot (1 - m_t)$ is the inferred part. Since $x_t \odot m_t = x_t$, Eq. (9.30) can be written as

$$\tilde{x}_t = x_t + \tilde{x}_t \odot (1 - m_t). \quad (9.31)$$

Since the transition of completed states follows the graph Markov process, the GMP with respect to the completed state can be described as $\tilde{x}_{t+1} =$

$\gamma (\mathbf{A} \odot P_t)\tilde{x}_t$. For simplicity, we denote the graph Markov transition matrix by H_t, i.e., $H_t = \mathbf{A} \odot P_t$. Hence, the transition process of completed states can be represented as

$$\tilde{x}_{t+1} = \gamma H_t \tilde{x}_t. \tag{9.32}$$

Applying Eq. (9.31), the transition process becomes

$$\tilde{x}_{t+1} = \gamma H_t (x_t + \tilde{x}_t \odot (1 - m_t)). \tag{9.33}$$

If we iteratively apply the completed state $\tilde{x}_t$, i.e., $\tilde{x}_t = \gamma H_{t-1}(x_{t-1} + \tilde{x}_{t-1} \odot (1 - m_{t-1}))$, into Eq. (9.33) itself, we have

$$
\begin{aligned}
\tilde{x}_{t+1} &= \gamma H_t (x_t + \gamma H_{t-1}(x_{t-1} + \tilde{x}_{t-1} \odot (1 - m_{t-1})) \odot (1 - m_t)) \\
&= \gamma H_t x_t + \gamma^2 H_t H_{t-1}(x_{t-1} \odot (1 - m_t)) \\
&\quad + \gamma^2 H_t H_{t-1}(\tilde{x}_{t-1} \odot (1 - m_{t-1}) \odot (1 - m_t)).
\end{aligned} \tag{9.34}
$$

After iteratively applying n steps of previous states from $x_{t-(n-1)}$ to x_t, $\tilde{x}_{t+1}$ can be described as

$$
\begin{aligned}
\tilde{x}_{t+1} &= \gamma H_t x_t \\
&\quad + \gamma^2 H_t H_{t-1}(x_{t-1} \odot (1 - m_t)) \\
&\quad + \gamma^3 H_t H_{t-1} H_{t-2}(x_{t-2} \odot (1 - m_{t-1}) \odot (1 - m_t)) + \cdots \\
&\quad + \gamma^n H_t \cdots H_{t-(n-1)}(x_{t-(n-1)} \odot (1 - m_{t-(n-2)}) \odot \cdots \odot (1 - m_t)) \\
&\quad + \gamma^n H_t \cdots H_{t-(n-1)}(\tilde{x}_{t-(n-1)} \odot (1 - m_{t-(n-1)}) \odot \cdots \odot (1 - m_t))
\end{aligned} \tag{9.35}
$$

$$
\begin{aligned}
&= \sum_{i=0}^{n-1} \gamma^{i+1} \left(\prod_{j=0}^{i} H_{t-j} \right)(x_{t-i} \odot \bigodot_{j=0}^{i-1}(1 - m_{t-j})) \\
&\quad + \gamma^n H_t \cdots H_{t-(n-1)}(\tilde{x}_{t-(n-1)} \odot (1 - m_{t-(n-1)}) \odot \cdots \odot (1 - m_t)),
\end{aligned} \tag{9.36}
$$

where $\sum$, $\prod$, and $\odot$ are the summation, matrix product, and Hadamard product operators, respectively. For simplicity, we denote the term with the $\tilde{x}_{t-n}$ in Eq. (9.35) as $\mathcal{O}(\tilde{x}_{t-n})$, and the GMP of the complected states can be represented by

$$\tilde{x}_{t+1} = \sum_{i=0}^{n-1} \gamma^{i+1} \left(\prod_{j=0}^{i} H_{t-j} \right)(x_{t-i} \odot \bigodot_{j=0}^{i-1}(1 - m_{t-j})) + \mathcal{O}(\tilde{x}_{t-(n-1)}). \tag{9.37}$$

In $\mathcal{O}(\tilde{x}_{t-(n-1)})$, when $n \to \infty$, since $\gamma \in (0, 1)$, $\gamma^{n+1} \to 0$. In addition, the product of masking vectors in $\mathcal{O}(\tilde{x}_{t-(n-1)})$ will also approach to zero, i.e.,

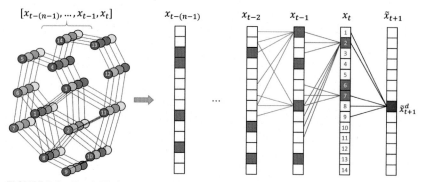

FIGURE 9.6 Graph Markov process. The gray-colored nodes in the left-hand sub-figure demonstrate the nodes with missing values. Vectors on the right-hand side represent the traffic states. The traffic states at time t are numbered to match the graph and the vector. The future state (in red color) can be inferred from their neighbors in the previous steps.

$\bigodot_{j=0}^{n-1}(1 - m_{t-j}) \to 0$. Fig. 9.6 demonstrates the graph Markov process for inferring the future state. The traffic network graphs with attribute-missed nodes (in gray color) is converted into traffic state vectors. The inference of $\tilde{x}_{t+1}^d$ is based on historical traffic states by back-propagating to the $t - (n - 1)$ step.

9.4.2 Method: graph Markov network

In this section, we propose a **Graph Markov Network** (GMN) for traffic prediction with the capability of handling missing values in historical data. Suppose the historical traffic data consists of n steps of traffic states $\{x_{t-(n-1)}, ..., x_t\}$. Correspondingly, we can acquire n masking vectors $\{m_{t-(n-1)}, ..., m_t\}$. The traffic network's topological structure can be represented by the adjacency matrix.

The GMN is designed based on the proposed GMP described in the previous section. Since we consider the term $\mathcal{O}(\tilde{x}_{t-(n-1)})$ described in Eq. (9.37) is small enough, the $\mathcal{O}(\tilde{x}_{t-(n-1)})$ is omitted in the proposed GMN for simplicity.

As described in Eq. (9.37), the graph Markov process contains n transition weight matrices, and the product of the these matrices ($\prod_{j=0}^{i} H_{t-j}$) = ($\prod_{j=0}^{i} \mathbf{A}^j \odot P_{t-j}$) measures the contribution of x_{t-i} for generating the $\tilde{x}_t$. To reduce matrix product operations and at the same time keep the learning capability in the GMP, we simplify the ($\prod_{j=0}^{i} \mathbf{A}^j \odot P_{t-j}$) by ($\mathbf{A}^{i+1} \odot W_{i+1}$), where $W_{i+1} \in \mathbb{R}^{S \times S}$ is a weight matrix. In this way, ($\mathbf{A}^{i+1} \odot W_{i+1}$) can directly measure the contribution of x_{t-i} for generating the $\tilde{x}_t$ and skip the intermediate state transition processes. Further, the GMP still has n weight matrices ($\{\mathbf{A}^1 \odot W_1, ..., \mathbf{A}^n \odot W_n\}$), and thus, the learning capability in terms of the size of parameters does not change. The benefits of the simplification is that the GMP can reduce $\frac{n(n-1)}{2}$ times of multiplication between two $S \times S$ matrices in total.

Based on the GMP and the aforementioned simplification, we propose the ***graph Markov network*** for traffic forecasting with the capability of handling

⊛ : Matrix product
⊙ : Hadamard product
⊕ : Sum

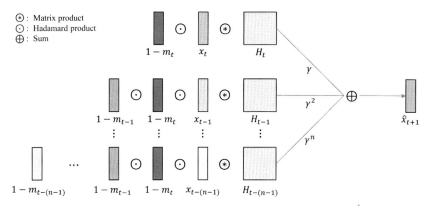

FIGURE 9.7 Structure of the proposed graph Markov network. Here, $H_{t-j} = \mathbf{A}^j \odot W_j$. As for the spectral version of GMN, $H_{t-j} = U \Lambda_j U^T$.

missing values as

$$\hat{x}_{t+1} = \sum_{i=0}^{n-1} \gamma^{i+1} (\mathbf{A}^{i+1} \odot W_{i+1})(x_{t-i} \odot \bigodot_{j=0}^{i-1}(1 - m_{t-j})), \qquad (9.38)$$

where $\hat{x}_{t+1}$ is the predicted traffic state for the future time step $t + 1$ and $\{W_1, ..., W_n\}$ are the model's weight matrices that can be learned and updated during the training process. The structure of GMN for predicting traffic state x_{t+1} is demonstrated in Fig. 9.7. During the training process, the loss can be calculated by measuring the difference between the predicted value $\hat{y} = \hat{x}_{t+1}$, and the label $y = x_{t+1}$.

9.4.3 Results

We compared the proposed approach with state-of-the-art traffic forecasting models with the capability of handling missing values.

Datasets: In this study, we conduct experiments on a real-world network-wide traffic speed dataset. The topological structure of the traffic network is also used in the experiments. This dataset named as PEMS-BAY is collected by California Transportation Agencies (CalTrans) Performance Measurement System (PeMS). This dataset contains the speed information of 325 sensors in the Bay Area, spanning the six months from Jan 1, 2017 to Jun 30, 2017. The interval of time steps is 5-minutes. The total number of observed traffic data points is 16,941,600. The dataset is published by Li et al. (2018) on the Github (https://github.com/liyaguang/DCRNN).

Missing Values and Data formatting: The dataset forms as a spatiotemporal matrix, whose spatial dimension size is the number of sensors and temporal dimension size is the number of time steps. In the experiments, the dataset is split into a training set, a validation set, and a testing set, with a size ratio of 6 : 2 : 2. In the training and testing process, the speed values of the dataset are normalized within a range of [0, 1]. There are no missing values in the PEMS-BAY datasets. To test the model's capability of handling missing values with various missing rates, we randomly set a portion of speed values in the input sequences to zeros according to a specific missing rate and generate the masking vectors accordingly. In this study, three groups of experiments are conducted, whose missing rates are 0.2, 0.4, and 0.6, respectively.

Baseline Models:

- GRU (Cho et al., 2014): GRU referring to gated recurrent units is a type of RNN. GRU can be considered as a simplified LSTM.
- GRU-I : GRU-I is designed based on GRU. Since GRU is a type of a RNN with the recurrent structure, the predicted values from a previous step $\hat{x}_t$ can be used to infer the missing values in the next step. The complected state can be represented by $\tilde{x}_{t+1} = x_t + \hat{x}_t \odot (1 - m_t)$.
- GRU-D (Che et al., 2018): GRU-D a neural network structure is designed based on GRU for multivariate time series prediction. It can capture long-term temporal dependencies in time series. GRU-D incorporates the masking and time interval into the model's architecture such that it can utilize the missing patterns.
- LSTM (Hochreiter and Schmidhuber, 1997): LSTM is a powerful variant of RNN that can overcome the gradients exploding or vanishing problem. It is suitable for being a model's basic structure for traffic forecasting.
- LSTM-I : LSTM-I is designed based on LSTM. The missing value inferring process of LSTM-I is similar to that of GRU-I.
- LSTM-M (Tian et al., 2018): LSTM-M is a neural network structure whose design is based on LSTM for traffic prediction with missing data. It employs multi-scale temporal smoothing to infer lost data.

Model Parameters The batch size of the tested data is set as 64. The number of steps of historical data incorporated in the GMN will have an influence on the prediction performance. Hence,the GMNs with 6-steps, 8-steps, and 10-steps of historical data are tested in the experiments, i.e., the n in Eq. (9.29) is set as 6, 8, and 10, respectively. In this experiment's section, we denoted these models by GMN-6, GMN-8, and GMN-10, respectively. The decay parameter γ in GMN is set as 0.9 in the experiments. For the RNN-based baseline models, including GRU-I, GRU-D, LSTM-I, and LSTM-M, the size of the input sequence is set as 10. In the training process, the mean square error (MSE) between the label y_t and the predicted value $\hat{y}_t$, i.e., $\frac{1}{n}\sum_{i=1}^{N}(y_i - \hat{y}_i)^2$ is used as the loss function, where N is the sample size. The Adam optimization method is adopted for both GMN models and baseline models.

Evaluation Metric: The prediction accuracy of all tested models are evaluated by three metrics, including mean absolute error (MAE), mean absolute percentage error (MAPE), and root mean square error (RMSE):

$$MAE = \frac{1}{N} \sum_{i=1}^{N} |y_i - \hat{y}_i| \qquad (9.39)$$

$$MAPE = \frac{1}{N} \sum_{i=1}^{N} |\frac{y_i - \hat{y}_i}{y_i}| \qquad (9.40)$$

$$RMSE = \left(\frac{1}{N} \sum_{i=1}^{N} |y_i - \hat{y}_i|^2 \right)^{\frac{1}{2}}. \qquad (9.41)$$

The prediction results of the compared models with respect to various missing rates are displayed in Table 9.1. When the missing rate is 0.1, GMN-10 achieves the smallest MAE and MAPE. However, the RMSEs of GRU-I and LSTM-M are less than those of GMN models. For the case in which the missing rate is set as 0.2, GMN-6 achieves the best prediction performance in terms of MAE and MAPE. Similarly, GRU-I's RMSE is smaller. When the missing rate is as large as 0.4, the MAE and MAPE results of the LSTM-M and GMN models are close, but the RMSEs of GMN models are more significant than that of LSTM-M. The overall prediction accuracy of GMN models in terms of MAE and MAPE is better than the baseline models. The RMSEs of GMN models is slightly larger than some of the compared models. Since the errors in RMSE are squared before they are averaged, the RMSE gives a relatively high weight to large errors. That means some of the predicted values generated by GMN models may have more significant errors.

9.5 Case study 3: graph neural network (GNN) for vehicle keypoints' correction

9.5.1 Problem definition

With the development of intelligent transportation system (ITS), surveillance cameras are becoming one of the most widely used traffic sensors worldwide for traffic data collection, monitoring, and management. Sensing 3-D structural information of vehicles via the monocular surveillance camera is very important for vehicle's real-time high precision positioning and scene reconstruction. But it is a challenging task. This case study treats a 3-D structural information sensing system that is supported by a customized fully-automatic camera calibration, for real-time vehicle localization and scene reconstruction based on a single monocular surveillance camera. It basically has three steps: 1) pre-processing module for vehicle detection and segmentation; 2) detecting 2-D car keypoints; and 3) projecting 2-D keypoints into 3-D coordinates and fulfilling the fully automatic camera calibration. We will demonstrate how the graph neural network can correct the detected keypoints.

TABLE 9.1 Prediction performance.

Model	Missing Rate = 0.1			Missing Rate = 0.2			Missing Rate = 0.4		
	MAE	MAPE	RMSE	MAE	MAPE	RMSE	MAE	MAPE	RMSE
GRU	1.594	3.097%	2.589	1.802	3.545%	2.93	2.253	4.524%	3.651
GRU-I	1.109	2.127%	**1.833**	1.185	2.29%	**1.968**	1.437	2.841%	2.412
GRU-D	5.21	13.28%	9.052	5.326	13.4%	9.042	5.242	13.33%	9.063
LSTM	2.372	4.836%	3.839	2.318	4.781%	3.888	2.469	5.185%	4.234
LSTM-I	3.129	5.418%	11.372	1.902	3.5%	6.638	1.965	3.793%	5.64
LSTM-M	1.18	2.307%	1.942	1.24	2.443%	2.066	**1.34**	2.665%	**2.297**
GMN-6	1.062	2.036%	2.321	**1.092**	**2.122%**	2.024	1.811	3.607%	3.534
GMN-8	1.066	2.042%	2.356	1.188	2.275%	2.739	1.404	2.771%	2.694
GMN-10	**1.062**	**2.033%**	2.365	1.201	2.296%	2.83	**1.34**	**2.636%**	2.458

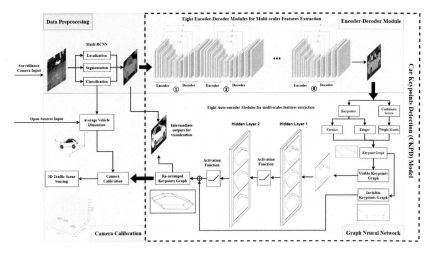

FIGURE 9.8   3-D Structural Information Sensing System (3D-SISS) Architecture.

The entire architecture of the 3-D structural information sensing system (3D-SISS) is shown in Fig. 9.8. The system can be divided into three parts: data preprocessing, car keypoints detection (CKPD) model, and 3D structural information projection.

Since 3D-SISS needs more detailed and accurate vehicle classification results to match the average vehicle dimension for camera calibration, this case study retrained Mask-RCNN with MIO-TCD dataset (Luo et al., 2018), a dataset containing eleven particular categories of target transportation agents in surveillance camera view.

To realize highly precise car pose and orientation estimation, we use the extracted vehicle segmentation from Mask-RCNN model to generate 12 predefined car keypoints in image coordinate, as shown at the right top corner in Fig. 9.8. The 3-D scene reconstruction through camera calibration method has two modules: 1) an encoder-decoder module for car keypoints detection; and 2) a GNN model for keypoints' correction.

In the encoder–decoder module, the research team proposes an end-to-end architecture to connect eight fourth-order encoder–decoder networks for car keypoints' estimation. In car keypoints detection, there are always some keypoints are completely invisible in any point of view. This brings significant challenges on both ground-truth data labeling and key features understanding. As a result, the encoder–decoder module can estimate only the position of the invisible car keypoints based on the detected locations of visible keypoints, which results in accumulated errors. And the errors in car keypoints detection can be amplified in the following process of camera calibration and 3-D scene reconstruction. Therefore, the keypoints graph $G = (v, \varepsilon)$ based on the output car

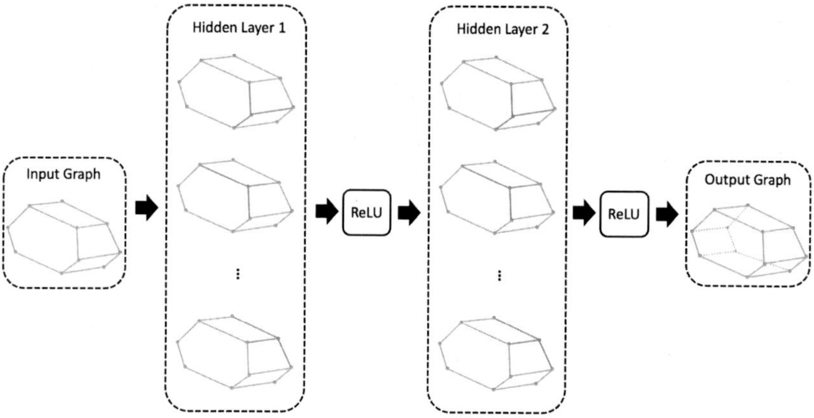

FIGURE 9.9 Structure of GNN for invisible car keypoints prediction.

keypoints is generated, and a GNN module is designed to extract the inter-keypoints relationships and correct the keypoints.

9.5.2 Method: graph neural network for keypoints correction

The GNN model used in 3D-SISS contains only two hidden layers to predict and correct the keypoints graph $G = (\upsilon, \varepsilon)$ with 12 keypoints and 18 edges. One thing that is worth mentioning here is that, before inputting graph G to the GNN, we remove the invisible keypoints and edges from G to increase the accuracy of the model. Based on the high- accuracy visible keypoints and edges, GNN can predict the invisible keypoints and edges for G. The output of GNN module is accurate 2-D car keypoints in image coordinates. The whole process is shown in Fig. 9.9.

First, keypoints graph $G = (V, E)$ with 12 keypoints are generated by the output of encoder–decoder network. In the process, we convert the heatmap into graph G by encoding the location and confidence score. Each keypoint υ_i in G can be represented as $\upsilon_i = \{x_i, y_i, w_i\}$, where (x_i, y_i) indicates the location of the keypoint in image coordinate and w_i is the weight of the keypoints in G, which can be represented as $w_i = c_i / \sum_i c_i, an$. Also, c_i indicates the confidence score of keypoint υ_i. Then, the relationship between all nodes is encoded in the edge $\epsilon_{ij} = \{\upsilon_i, \upsilon_j\}$, where

$$\epsilon_{ij} = \begin{cases} 1, & \forall i, j \in V \text{ and } i \neq j \\ 0, & \text{otherwise.} \end{cases} \qquad (9.42)$$

After defining the keypoint graph G, we will use the GNN model to predict the latent graph structure for keypoints' correction. Fig. 9.9 shows the structure of the GNN model. In the network, two hidden layers are used to predicted the invisible edges. The process is modeled as $q(\epsilon_{ij}|V) = softmax(f(V))$, where $f(V)$ is a GNN acting on the fully connected graph produced from the heatmap. The process can be represented by the Eq. (9.43). In Eq. (9.43), h^t indicates the tth layer of GNN; $\upsilon \to \epsilon$ indicates the convolutional operation from vertex to edge; $\epsilon \to \upsilon$ indicates the convolutional operation from edge to vertex; and f indicates the fully connected (FC) layer:

$$
\begin{cases}
\upsilon : h_i^1 = f(\upsilon_i) \\[2mm]
\upsilon \to \epsilon : h_{i,j}^1 = f^1([h_i^1, h_j^1]) \\[2mm]
\epsilon \to \upsilon : h_i^2 = f(\sum_{i \neq j} h_{i,j}^1) \\[2mm]
\upsilon \to \epsilon : h_{i,j}^2 = f^2([h_i^2, h_j^2]).
\end{cases}
\tag{9.43}
$$

In the model, our goal is to investigate the spatial relationship among keypoints, which is encoded as edges in graph G. Therefore, the edge loss is applied to the process to minimize the difference between predicted edges and the ground-truth edges.

9.5.3 Results

3D-SISS is a complex system which requires a comprehensive experiment to test its performance. Therefore, in the research, cooperating with the City of Bellevue and Connect Cities with Smart Transportation (C2SMART), we installed our own surveillance camera in the City of Bellevue testbed to test the 3D-SISS system. The location and the installation height is premeasured in the process for final results validation. We also adjusted the focus length and the angle of the surveillance camera thorough control panel. Therefore, we can get the ground-truth information about the camera on both internal and external sides for camera parameter estimation performance test.

The 3-D structural information sensing system can provide a detailed traffic status analysis without any human input or initialization. The car keypoints' detection model can achieve accurate (overall 95.1%) car keypoints' detection with single image input. As we tested at the testbed, the 3D-SISS can be applied to any camera at various view angles for real-time comprehensive traffic sensing, as shown in Fig. 9.10. By taking advantage of the most widely deployed surveillance camera without installing new sensors, 3D-SISS can expand the widely deployed cameras from the 2-D to 3-D view at a negligible cost.

FIGURE 9.10 Car keypoints detection results in various views.

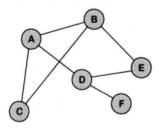

FIGURE 9.11 Example graph.

9.6 Exercises

9.6.1 Questions

1. Describe the difference between the adjacency matrix and the degree matrix of a graph.
2. Is GNN a type of supervised learning method or unsupervised learning method?
3. In Fig. 9.11, a graph is depicted; write down the adjacency matrix based on this figure.
4. We use a simplified spatial approach to fulfill the graph neural network layer based on Fig. 9.11, $Y = AWX$, where X is the nodes input value $X = [1, 2, 3, 4, 5, 6]^T$ and W is the weight matrix with all elements equaling to one. Write out the output Y.
5. In spatial-based GNNs, the equation may appear like $Y = \sigma(AWX)$, where σ is an activation function. Explain why it needs the σ function.
6. The adjacency matrix is the key to describe the topological structure for the data. Why do studies allow the diagonal elements of the adjacency matrix A be one in GNN ($\tilde{A} = A + I$)?
7. Explain the key difference between Graph Convolutional Neural Network (GCN) and Graph Attention Network (GAT).
8. In a graph dataset, if the degrees of all the nodes in the graph are equal to one, will you use a GCN to process the data? Explain the reason.
9. In a convolutional layer, the input and output may not have the same dimension. Is this the same for the graph convolutional layer? Explain the reason

and the difference between a convolutional layer and a graph convolutional layer.

10. List five transportation related scenarios that can apply GNN.

Chapter 10

Generative adversarial networks

Generative modeling is a type of machine learning where the aim is to model the distribution that a given set of data (e.g. images, audio) came from. Normally this is an unsupervised problem, in the sense that the models are trained on a large collection of data. When trained successfully, we can use the generative models to estimate the likelihood of each observation and to create new samples from the underlying distribution. For instance, recall that the MNIST dataset was obtained by scanning handwritten zip code digits from envelopes. So, consider the distribution of all the digits people ever write in zip codes. The training examples can be viewed as samples from this distribution. If we fit a generative model, we're trying to learn about the distribution from the training samples. Notice that this process doesn't use the labels, so it's an unsupervised learning problem. The reasons of train a generative model can be:

1. Most straightforwardly, we may want to generate realistic samples, e.g., for traffic image data-generation applications.
2. We may wish to model the data distribution in order to tell which of several candidate outputs is more likely.
3. We may want to train a generative model in order to learn useful high-level features that can be applied to other tasks.

There are various of deep generative models in widespread use today, such as generative adversarial networks, variational autoencoders, flow-based models, and diffusion models. Generative adversarial network, or GAN for short, is an unsupervised learning task in machine learning that involves automatically discovering and learning the regularities or patterns in input data in such a way that the model can be used to generate or output new examples that plausibly could have been drawn from the original dataset. GANs are a clever way of training a generative model by framing the problem as a supervised learning problem with two submodels: the **generator** model that we train to generate new examples, and the **discriminator** model that tries to classify examples as either real (from the domain) or fake (generated). The two models are trained together in a zero-sum game, adversarial, until the discriminator model is fooled about half the time, meaning the generator model is generating plausible examples.

In this chapter, we will introduce the structure and the formulation of GANs. In addition, we will also show several case studies to present the application scenarios of GANs.

Machine Learning for Transportation Research and Applications
https://doi.org/10.1016/B978-0-32-396126-4.00015-1

10.1 Generative adversarial network (GAN)

10.1.1 Binary classification

Given a data distribution $p_{\text{data}}(x, y)$ with $y \in \{0, 1\}$, we would like to fit a binary classifier $p_\phi(y \mid x)$ to the conditional distribution $p_{\text{data}}(y \mid x)$. A maximum likelihood estimate of the parameters ϕ is obtained by solving the following optimization task:

$$\phi^* = \arg\max_{\phi} \mathbb{E}_{p_{\text{data}}(x,y)} \left[\log p_\phi(y \mid x) \right]. \tag{10.1}$$

Assume the dataset is balanced, i.e., $p_{\text{data}}(y) = \text{Bern}(0.5)$, then the stated objective is equivalent to

$$\phi^* = \arg\max_{\phi} \mathbb{E}_{p_{\text{data}}(x|y=1)} \left[\log p_\phi(y = 1 \mid x) \right] \\ + \mathbb{E}_{p_{\text{data}}(x|y=0)} \left[\log \left(1 - p_\phi(y = 1 \mid x) \right) \right]. \tag{10.2}$$

The negation of the maximum likelihood objective is also known as the cross-entropy loss.

10.1.2 Original GAN formulation as binary classification

The generative adversarial network (GAN) approach (Goodfellow et al., 2020) constructs a binary classification task to assist the learning of the generative model distribution $p_\theta(x)$ to fit the data distribution $p_{\text{data}}(x)$. This is done by labeling all the datapoints sampled from the data distribution as "real" data and those sampled from the model as "fake" data. In other words, a joint distribution $\tilde{p}(x, y)$ is constructed as follows for the binary classification task:

$$\tilde{p}(x, y) = \tilde{p}(x \mid y)\tilde{p}(y), \qquad \tilde{p}(y) = \text{Bern}(0.5),$$

$$\tilde{p}(x \mid y) = \begin{cases} p_{\text{data}}(x), & y = 1 \\ p_\theta(x), & y = 0. \end{cases} \tag{10.3}$$

Fitting a binary classifier ("discriminator") with $p_\phi(y = 1 \mid x) = D_\phi(x)$ to $\tilde{p}(y \mid x)$ can be done by maximizing the maximum likelihood objective:

$$\phi^*(\theta) = \arg\max_{\phi} \mathcal{L}(\theta, \phi),$$

$$\mathcal{L}(\theta, \phi) := \mathbb{E}_{p_{\text{data}}(x)} \left[\log D_\phi(x) \right] + \mathbb{E}_{p_\theta(x)} \left[\log \left(1 - D_\phi(x) \right) \right]. \tag{10.4}$$

Notice the dependence of the objective on the generative model parameter θ since the "data distribution" $\tilde{p}(x, y)$ of the binary classification task depends on $p_\theta(x)$. Then, the training of the generative model $p_\theta(x)$ aims at fooling the discriminator, by minimizing the log probability of making the right decisions:

$$\theta^*(\phi) = \arg\min_{\theta} \mathbb{E}_{p_\theta(x)} \left[\log\left(1 - D_\phi(x)\right) \right].$$ (10.5)

In summary, the two-player game objective for training the GAN generator and discriminator is

$$\min_{\theta} \max_{\phi} \mathcal{L}(\theta, \phi).$$ (10.6)

GANs learns to map the simple latent distribution to the more complex data distribution. To capture the complex data distribution p_{data}, GANs architecture should have enough capacity. GANs is based on the concept of a noncooperative game of two networks, a generator G, and a discriminator D, in which G and D play against each other. GANs can be part of deep generative models or generative neural models, where G and D are parameterized via neural networks and updates are made in parameter space.

Both G and D play a minimax game, where G's main aim is to produce samples similar to the samples produced from real data distribution and D's main goal is to discriminate the samples generated by G and samples generated from the real data distribution by assigning higher and lower probabilities to samples from real data and generated by G, respectively. On the other hand, the main target of GANs training is to keep moving the generated samples in the direction of the real data manifolds through the use of the gradient information from D.

In GANs, x is data extracted from the real data distribution, p_{data}, the noise vector z is taken from a Gaussian prior distribution with zero-mean and unit variance p_z, while p_g refers the G's distribution over data x. Latent vector z is passed to G as an input and then G outputs an image $G(z)$ with the aim that D cannot differentiate between $G(z)$ and $D(x)$ data samples, i.e., $G(z)$ resembles with $D(x)$ as close as possible. In addition, D simultaneously tries to restrain itself from getting fooled by G. D is a classifier where $D(x) = 1$ if $x \sim p_{data}$ and $D(x) = 0$ if $x \sim p_g$, i.e., x is from p_{data} or from p_g.

10.1.3 Objective (loss) function

An objective function tries to match real data distribution p_{data} with p_g. Basic GANs use two objective functions: (1) D minimizes the negative log-likelihood for binary classification; (2) G maximizes the probability of generated samples for being real. D parameters are denoted by θ_D, which are trained to maximize the loss to distinguish between the real and fake samples. G parameters are denoted by θ_G, which are optimized such that the D is not able to distinguish between real and fake samples generated by G. θ_G is trained to minimize the same loss that θ_D is maximizing. Hence, it is a zero-sum game where players compete with each other. The following minimax objective applied for training

G and D models jointly via solving

$$\min_{\theta_G} \max_{\theta_D} V(G, D) = \min_{G} \max_{D} \mathbb{E}_{x \sim p_{\text{data}}} \big[\log D(x) \big] + \mathbb{E}_{z \sim p_z} \big[\log\big(1 - D\big(G(z)\big)\big) \big].$$
(10.7)

$V(G, D)$ is a binary cross entropy function, commonly used in binary classification problems. In Eq. (10.7), for updating the model parameters, training of G and D are performed by back-propagating the loss via their respective models. In practice, Eq. (10.7) is solved by alternating the following two gradient updates:

$$\theta_D^{t+1} = \theta_D^t + \lambda^t \nabla_{\theta_D} V\big(D^t, G^t\big) \quad \text{and} \quad \theta_G^{t+1} = \theta_G^t + \lambda^t \nabla_{\theta_G} V\big(D^{t+1}, G^t\big),$$
(10.8)

where θ_G is the parameter of G, θ_D is the parameter D, λ is the learning rate, and t is the iteration number. In practice, the second term in Eq. (10.7), $\log(1 - D(G(z)))$ saturates and makes insufficient gradient flow through G, i.e., the gradient's value gets smaller and stop learning. To overcome the vanishing gradient problem, the objective function in Eq. (10.7) is reframed into two separate objectives:

$$\max_{\theta_D} \mathbb{E}_{x \sim p_{\text{data}}} \big[\log D(x) \big] + \mathbb{E}_{z \sim p_z} \big[\log\big(1 - D\big(G(z)\big)\big) \big] \quad \text{and}$$
$$\max_{\theta_G} \mathbb{E}_{z \sim p_z} \big[\log\big(D\big(G(z)\big)\big) \big].$$
(10.9)

Moreover, G's gradient for these two separate objectives have the same fixed points and are always trained in the same direction. After the cost computation in Eq. (10.8), back-propagation can be used to update the model parameters. Because of these two different objectives, the update rule is given as

$$\big\{\theta_D^{t+1}, \theta_G^{t+1}\big\} \leftarrow \begin{cases} \text{Update} & \text{if } D(x) \text{ predicts wrong} \\ \text{Update} & \text{if } D(G(z)) \text{ predicts wrong} \\ \text{Update} & \text{if } D(G(z)) \text{ predicts correct.} \end{cases}$$
(10.10)

If D and G are given enough capability with sufficient training iterations, G can convert a simple latent distribution p_g into more complex distributions, i.e., p_g converges to p_{data}, such as $p_g = p_{\text{data}}$.

10.1.4 Optimization algorithm

In GANs, optimization is to find (global) equilibrium of the minmax game, i.e., the saddle point of minmax objective. Gradient-based optimization methods are widely used to find the local optima for classical minimization and saddle point problems. Any traditional gradient-based optimization technique can be used for minimizing each player's cost simultaneously which leads to the Nash equilibrium. Basic GANs uses the simultaneous gradient descent for finding the Nash equilibrium and update D's and G's parameters by simultaneously using gradient descent on D's and G's utility functions. Each player D and G tries to minimize its own cost/objective function for finding a Nash equilibrium (θ_D, θ_G), $J_D(\theta_D, \theta_G)$ for the D and $J_G(\theta_D, \theta_G)$ for the G, i.e., J_D is at a minimum w.r.t. θ_D and J_G is at a minimum w.r.t. θ_G.

Minimax optimization for Nash equilibrium. The total cost of all players in a zero-sum game is always zero. In addition, a zero-sum game is also known as a minimax game because solution includes minimization and maximization in an outer and inner loop, respectively. Therefore, i.e.,

$$J_G + J_D = 0$$
$$J_G = -J_D.$$

All the GANs game designed earlier apply the same cost function to the D, J_D, but they vary in G's cost, J_G. In these cases, D uses the same optimal strategy. D's cost function is as follows w.r.t. θ_D:

$$J_D(\theta_D, \theta_G) = -\frac{1}{2}\mathbb{E}_{x \rightsquigarrow p_{\text{data}}}\left[\log D(x)\right] - \frac{1}{2}\mathbb{E}_z\left[\log\left(1 - D\left(G(z)\right)\right)\right]. \quad (10.11)$$

Eq. (10.11) represents the standard cross-entropy cost, which is minimized during the training of a standard binary classifier with a sigmoid output. In the minimax game, G attempts to minimize and maximize the log probability of D being correct and being mistaken, respectively. In GANs, training of a classifier is performed on two minibatches of data: a real data minibatch having examples' label 1 and another minibatch from G having examples' label 0, i.e., the density ratio between true and generated data distribution is represented as follows:

$$D(x) = \frac{p_{\text{data}}(x)}{p_g(x)} = \frac{p(x \mid y = 1)}{p(x \mid y = 0)} = \frac{p(y = 1 \mid x)}{p(y = 0 \mid x)} = \frac{D^*(x)}{1 - D^*(x)}.$$

Here, $y = 0$ means generated data and $y = 1$ means real data. $p(y = 1) = p(y = 0)$ is assumed.

Through training D, we get an estimate of the ratio $p_{\text{data}}(x)/p_g(x)$ at every point x. D learns to distinguish samples from data for any given G. The optimal D for a fixed G is given by $D_G^*(x) = \frac{p_{\text{data}}(x)}{p_{\text{data}}(x) + p_g(x)}$. After sufficient training steps, G and D with enough capacity will converge to $p_g = p_{\text{data}}$, i.e., D cannot discriminate between two distributions. For optimal D, the minimax game in Eq. (10.7) can now be reformulated as follow:

$$J_D = -2\log 2 + 2\left(D_{JS}\left(p_{\text{data}} \| p_g\right)\right), \quad (10.12)$$

where D_{JS} denotes Jensen–Shannon divergence. When D is optimal, G minimizes Jensen–Shannon divergence (JSD) mentioned in Eq. (10.12) which is an alternative similarity measure between two probability distributions, bounded by [0, 1]. In the basic GANs, it is feasible to estimate neural samplers through approximate minimization of the symmetric JSD.

$$D_{JS}\left(p_{\text{data}} \| p_g\right) = \frac{1}{2} D_{KL}\left(p_{\text{data}} \| \frac{1}{2}\left(p_{\text{data}} + p_g\right)\right)$$
$$+ \frac{1}{2} D_{KL}\left(p_g \| \frac{1}{2}\left(p_{\text{data}} + p_g\right)\right),$$

where D_{KL} denotes the KL divergence.

JSD is based on KL divergence, but it is symmetric, and it always has a finite value. Because $D_{JS}\left(p_{\text{data}} \| p_g\right)$ is an appropriate divergence measure between distributions, i.e., the real data distribution p_{data} can be approximated properly when sufficient training samples exist and model class p_g can represent p_{data}. The JSD between two distributions is always nonnegative and zero when two distributions are equivalent, $J_D^* = -2 \log 2$ is the global minimum of J_D which shows $p_g = p_{\text{data}}$.

Given the minimax objective, $p_g = p_{\text{data}}$ occurs at a saddle point θ_G, θ_D. The saddle point of a loss function occurs at a point that is minimal w.r.t. one set of weights and maximal w.r.t. another. GANs training interchanges between minimization and maximization steps. Thus, finding Nash equilibrium in GANs is a very challenging problem because objective functions are non-convex, parameters are continuous, and the parameter space is high-dimensional, e.g., an update to θ_D that decreases J_D can increase J_G and an update to θ_G that decreases J_G can increase J_D, i.e., training fails to converge. GANs are still hard to train and training suffers from the following problems, such as difficulty converging and instability, and mode collapse. A lot of techniques and GAN variants can help solve these problem, but they will not be discussed in this chapter due to the space limits.

10.2   Case studies: GAN-based roadway traffic state estimation

Traffic state information plays an important role in relieving traffic congestion. Accurate traffic state information can help administrators effectively manage the transportation network and travelers better plan their trips. Existing machine learning methods have contributed a lot to traffic estimation by adopting new datasets or designing novel methodologies. Traffic flow parameters in a transportation network have correlation over both time and space according to the traffic flow theory. Thus, incorporating traffic flow theory is beneficial for improving the accuracy of traffic state estimation. However, most of the learning-based models in the literature fail to do so when estimating a traffic state. Thus, as Fig. 10.1 shows, this case study details attempts to present how GAN can help the process of generating/estimating traffic along a road/corridor, while incorporating traffic flow theory.

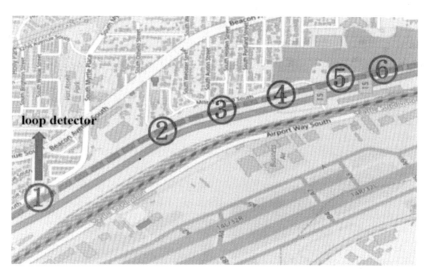

FIGURE 10.1 Estimating traffic states on the freeway segment on I-5 corridor in Seattle, WA.

10.2.1 Problem formulation

Traffic flow theory, such as the cell transmission model Daganzo (1994), indicates that traffic flow, traffic density, and traffic velocity in a transportation network are correlated. Further, these correlations exist in both time and space. Incorporating these spatiotemporal correlations can improve the accuracy of traffic state estimation. For example, studies has attempted to input traffic flow and traffic velocity to LSTM models to improve traffic velocity prediction. Spatiotemporal correlations among traffic flow parameters are also incorporated in traffic state estimation. However, in those studies, spatiotemporal correlations among traffic flow parameters are only mined from data, and no specific theoretical correlations are defined due to their relatively fixed model structures. GANs are able to incorporate correlations among traffic flow parameters from both the data and theory due to their flexible framework.

The traffic state-estimation problem addressed in this case study can be described as follows. We choose traffic flow and traffic density as inputs to our proposed model. As traffic flow and traffic density evolve with time and space, both traffic flow and traffic density within a given spatiotemporal area can be represented as a 2-D matrix. We assume the number of time intervals for the study period is n and that the road segment can be divided into m discrete cells. The $n*(m+1)$ traffic flow matrix is denoted as F with each element $q(t, l)$ representing the traffic flow at location l at time t. The $n*m$ traffic density matrix is denoted as K with each element $k(t, s)$ representing the traffic density in cell s at time t. The problem addressed in this study is how to estimate or predict the missing values in F and K, given known elements in each matrix. Note that the proposed model can be viewed as a deep-learning-based and data-driven cell transmission model if the known elements are boundary conditions.

10.2.2 Model: generative adversarial architecture for spatiotemporal traffic-state estimation

In this case study, the proposed methodology for traffic state estimation utilizes the G and D networks (both are LSTM networks), pretrained with uncorrupted data, to reconstruct traffic state. Denote the corrupted traffic state matrix as $\mathbf{y}$. We do not use D to update $\mathbf{y}$ by maximizing $D(\mathbf{y})$. Similar to the image inpainting (Yeh et al., 2016), maximizing $D(\mathbf{y})$ does not lead to the desired reconstruction. This is mainly because $\mathbf{y}$ is neither on the p_{data} manifold, nor on the G manifold, and the corrupted data is not drawn from those distributions. Therefore, we consider using both G and D for reconstruction. To quantify the 'closest' mapping from $\mathbf{y}$ to the reconstruction, we define three loss functions: a contextual loss, a perceptual loss, and a conservative loss.

Contextual loss: We need to incorporate the information from the known information of the given corrupted traffic state matrix into the traffic state estimation. The contextual loss is used to measure the context similarity between the reconstructed traffic state matrix and the uncorrupted portion. The contextual loss is defined as

$$L_{\text{contextual}}\,(\mathbf{z}) = \|M \odot G(\mathbf{z}) - M \odot \mathbf{y}\|_1, \qquad (10.13)$$

where M is a binary matrix having the same dimension as the output traffic matrices of G, denoting the mask of the corrupted traffic matrix. Elements equal to one in M indicate the corresponding element in the corrupted traffic matrix is uncorrupted. Elements equal to zero in M indicate the corresponding element in the corrupted traffic matrix is missing. $\odot$ denotes the element-wise product operation. The corrupted portion, i.e., $(1 - M) \odot \mathbf{y}$ is not used in the loss. The choice of the ℓ_1-norm is empirical.

Perceptual loss: The perceptual loss encourages the reconstructed traffic state matrix to be similar to the samples drawn from the training set. This is achieved by updating $\mathbf{z}$ to fool D. As a result, D will predict $G(\mathbf{z})$ to be from the training data with a high probability. We use the same loss for fooling D as in GANs:

$$L_{\text{perceptual}}\,(\mathbf{z}) = \log\bigl(1 - D\bigl(G(\mathbf{z})\bigr)\bigr). \qquad (10.14)$$

Without $L_{\text{perceptual}}$, some reconstructed traffic state matrices tend to be unrealistic.

Conservative loss: According to the conservation law of traffic flow, the spatiotemporal relationship between traffic volume and traffic density can be described as

$$k(t+1,\ s) = k(t, s) + \frac{\Delta t}{\Delta x_s}\bigl(q_{\text{in}}\,(t, s) - q_{\text{out}}\,(t, s)\bigr), \qquad (10.15)$$

where $k(t, s)$ is the vehicle density of cell s at time index $t, q_{in}(t, s)$ and $q_{out}(t, s)$ are the total flows (in vehicles per unit time) entering and leaving cells during the time interval $[t \cdot \Delta t, (t + 1) \cdot \Delta t]$, respectively. Δt is the sampling duration, and Δx_s is the length of cell s. In the context of traffic state estimation using GAA, the conservative loss is defined as

$$K_{G(z)}(t + 1, s) = K_{G(z)}(t, s) + \frac{\Delta t}{\Delta x_s} \left(F_{G(z)}(t, s) - F_{G(z)}(t, s + 1) \right), \quad (10.16)$$

where $K_{G(z)}$ and $F_{G(z)}$ are the traffic density matrix and the traffic flow matrix generated by the pretrained G, respectively. Eq. (10.16) describes the theoretical spatiotemporal correlation among traffic flow and traffic density. Thus, it helps to improve the accuracy of traffic state estimation.

With the defined loss functions, the traffic state matrix with missing values can be mapped to the closest $\mathbf{Z}$ in the latent representation space. $\mathbf{z}$ is updated using back-propagation with the total loss of

$$\hat{\mathbf{z}} = \arg \min_{\mathbf{z}} \left(L_{contextual}(\mathbf{z}) + \lambda_p L_{perceptual}(\mathbf{z}) + \lambda_c L_{conservative}(\mathbf{z}) \right), \quad (10.17)$$

where λ_p and λ_c are weighting parameters. In practice, the weighting parameters have to be relatively small to ensure the traffic state-estimation accuracy. After finding $\hat{\mathbf{z}}$, the estimated traffic state can be obtained by

$$\mathbf{x}_{reconstructed} = \mathbf{M} \odot \mathbf{y} + (1 - \mathbf{M}) \odot G(\hat{\mathbf{z}}). \quad (10.18)$$

At this stage, the GAN-based model for roadway traffic state estimation is established.

10.2.3   Results

This case study proposes a generative adversarial architecture (GAA) with an LSTM discriminator and an LSTM generator both containing 16 hidden neurons in the hidden layer. The proposed generative adversarial architecture for spatiotemporal traffic state estimation is evaluated on the loop data collected from a segment of the I-5 freeway in Seattle as shown in Fig. 10.1. The segment of interest of the I-5 freeway has six loop detectors installed along it. The data is collected every 20 sec covering the whole year of 2015.

Each training data record in the experiment consists of data across one hour from the six loop detectors. In the evaluation, two-thirds of the data are randomly selected for training, and the rest is used for validation.

Four types of GAAs are evaluated, including a GAA without perceptual loss and conservative loss (GAA model 1); a GAA without perceptual loss (GAA model 2); a GAA without conservative loss (GAA model 3); and a GAA with

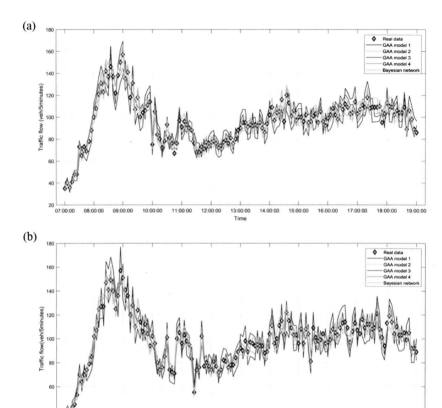

FIGURE 10.2   Model comparison result.

perceptual loss and conservative loss (GAA model 4). The four types of GAAs are compared with the existing Bayesian network approach proposed by (Sun et al., 2006) in the experiment.

Each record in the testing dataset is a pair of average traffic flow and traffic density measurements computed over one hour. We used the data in the previous half hour (i.e., 6*5 minutes) to predict traffic flow and traffic density in the next half hour (i.e., the next 6*5 minutes). Fig. 10.2 shows the traffic state estimated by the five models on the dataset. These figures shows the traffic flow estimated by the five models from 7:00AM to 7:00PM on Oct. 10, 2015. In the traffic flow estimation, the difference between the GAA without conservative loss and the GAA without perceptual loss is marginal in terms of MAPE, with only 0.08% MAPE difference, while, in other cases, the GAA without conservative loss outperforms the GAA without perceptual loss.

10.3 Case study: conditional GAN-based taxi hotspots prediction

It is of practical importance to predict taxi hotspots in urban traffic networks. However, the identification of taxi hotspot is generally influenced by multiple sources of dependence. To address this problem, this case study (Yu et al., 2020) was designed to investigate how the integration of clustering models and deep learning approaches can learn and extract the network-wide taxi hotspots in both temporal and spatial dimensions. A conditional generative adversarial network with a long short-term memory structure (LSTM-CGAN) model was proposed for taxi hotspot prediction, which is capable of capturing the spatial and temporal variations of the taxi hotspots in the same time. The proposed LSTM–CGAN model was then trained by the network-wide hotspot data. The comparative analyses suggest that the proposed LSTM–CGAN model outperformed other benchmark methods and demonstrate great potential to enable many shared mobility applications.

10.3.1 Problem formulation

The emergence of deep learning approaches has the potential to advance the understanding of taxi-based mobility demands and to enable the network-wide short-term taxi hotspot prediction. The primary objective of this case study is to develop a network-wide taxi hotspot prediction framework via an integration model of clustering and deep learning approaches. To solve the hotspot prediction problem, a spatiotemporal equivalent distance is applied in the a density-based clustering approach to cluster geocoded taxi pickup objects into a list of spatiotemporal taxi hotspot indicators, so that spatial distribution and possible temporal correlation of homogeneous mobility characteristics is naturally captured. Inspired by Liang et al. (2018), a adversarial architecture, namely the conditional generative adversarial network with long short-term memory structure (LSTM-CGAN), is modeled to capture the spatiotemporal distribution of taxi hotspots using the clustering results and corresponding information.

10.3.2 Model: LSTM-CGAN-based-hotspot prediction

An overview the taxi hotspot extraction and prediction framework is illustrated in Fig. 10.3. The network-wide taxi-hotspot extraction and prediction framework is developed by integrating a density-based spatiotemporal clustering algorithm with noise (DBSTCAN) and a conditional generative adversarial network with long short-term memory structure (LSTM–CGAN). The DBSTCAN algorithm was applied to capture the spatiotemporal correlation of taxi pickups by applying the spatiotemporal distance.

The basic adversarial architecture in the LSTM–CGAN model consists of the generative structure G and the discriminate structure D. The proposed LSTM–CGAN model enables the GAN model to cope with conditional information and

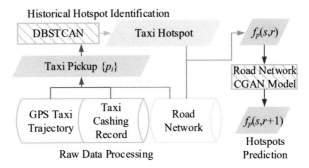

FIGURE 10.3 Framework overview for the spatiotemporal hotspot identification and prediction.

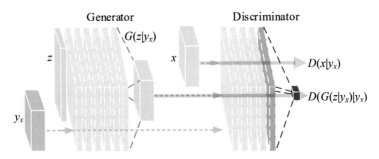

FIGURE 10.4 Conditional GAN in the proposed LSTM–CGAN model.

LSTM technique. More specifically, the generator G captures the data distribution, and the discriminator D estimates the probability that a sample come from the true data rather than generative network G. For data requirement, the adversarial architecture are trained in a semi-supervised way, which requires no additional label information. For output results, the generative adversarial network is able to approximating intractable probabilities, while other approaches tends to learn the average pattern.

In this case study, considering the fact that the location of taxi hotspots is influenced by various types of dependencies, conditional information is applied in both generator G and discriminator D as an additional input layer. With the conditional information, the generator G is able to specify the most probable outputs within the underlying distribution. The conditional structure applied in the proposed LSTM–CGAN model is illustrated in Fig. 10.4. Generator G builds a mapping function from a prior noise distribution p_z and the conditional information y to a data space, termed $G(z \mid y)$. At the same time, the output $D(x \mid y_x)$ from discriminator D scales the probability that input data x comes from true data rather than G. Both G and D are trained simultaneously. The parameters in generator G are adjusted to fool the D, while the parameters in D are estimated to maximize the capability of identifying both the true data and samples output from G. In other words, generator G and discriminator D play a two-player minimax game, where G is updated to maximize the similarity between random samples and true data.

Furthermore, considering the temporal dependencies of taxi hotspots, two LSTM neural networks are utilized in the G and D, respectively. The LSTM neural network has an benefit for dealing with long-term memory in temporal issues. A LSTM cell maps the input vector sequence $\{x, y_x\}$ to an output vector sequence h_t by T_0 iterations. In the proposed LSTM–CGAN structures, two LSTM neural networks are utilized. In the sample layout, shown in Fig. 10.4, a generator G consists of seven layers, which are one random generation layer, and six LSTM layers; and a discriminator D also consists of four LSTM layers, one fully connected layer and one softmax layer. In this case study, different cell numbers in a single LSTM layer were tested. In order to prevent over-fitting, we further apply the dropout method with a probability of 0.7 at each LSTM layer, which means that 30% cells are randomly ignored in each layer. The input of the D is designed to include a sequence of historical hotspot x_r for each temporal unit and the corresponding conditional information y_{x_r}, i.e., the taxi hotspot information in previous temporal unit and the time-of-day information.

10.3.3   Results

The taxi trajectory data and taximeter data used in this paper was collected from Beijing taxi companies. The number of referred taxis was 68,310. As shown in the annual city report of Beijing, the total number of operating taxis was around 68,000, which indicated that the authors obtained all the accessible taxi data during the study period. The total number of pick-up records in the study area was 2,640,826. The study period was two weeks (from Nov. 2, 2015 to Nov. 15, 2015). More specifically, data from Nov. 2 to Nov. 12 was used as the training set, data form Nov. 13 was used as a validating set, and the other two days were used for testing. The major purpose of this study was to develop the extraction and prediction framework for taxi hotspots and examined its feasibility and predictive performance under prevailing conditions. Two week traffic data under prevailing conditions is sufficient to demonstrate major, regular traffic-operation patterns for the proposed method to capture the spatiotemporal features in terms of preliminary testing purposes. Fig. 10.5 shows the spatial distribution of taxi pickups in various time periods for both weekday.

In order to better understand the superiority of the proposed method against the reference methods, especially for the unbalanced dataset, the receiver operating characteristic (ROC) curve is employed, and the performance of difference approaches is measured by the area under the curve (AUC). More specifically, the ROC curve is a graphical plot that illustrates the prediction performance of the proposed model as the discrimination threshold is varied. It is created by plotting the true positive rate (1-FIT-N), against the false positive rate (FIT-P), and the area under the receiver operating characteristic (AUROC). Fig. 10.6 illustrated the ROC curves for the prediction performance of the five methods.

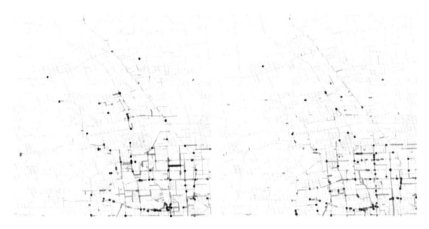

FIGURE 10.5 Spatial distribution of taxi pickups in various time periods.

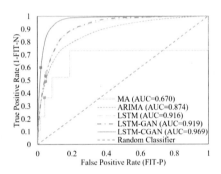

FIGURE 10.6 ROC curves for the prediction performance considering various thresholds.

Note that the prediction performance using 0.5 as the threshold is labeled with a solid block on each curve. As shown in Fig. 10.6, almost all five methods perform better than the random classifier, except for the MA method when the threshold is very low. It was found that the proposed LSTM–CGAN outperformed all the other methods, followed by the LSTM–GAN model, the LSTM model, the ARIMA model, and the MA method, which are all baseline methods.

10.4 Case study: GAN-based pavement image data transferring

Pavement distress significantly endangers traffic safety and reduces road life. Automated pavement damage-detection technology plays a pivotal role in the rapid evaluation of road conditions and maintenance. Images have become the primary data source for pavement distress detection due to their accessible collection and available sophisticated processing methods. A wide variety of devices are used for data collection, such as driving recorders, car cameras, and unmanned aerial vehicles. This diversity of application scenarios makes the styles and elements of images quite distinct. Due to the different placements and angles of the collection equipment, apparent differences lie in those images in terms of the complexity of the background elements, the degree of inclination, and distortion of the pavement distress. Thus, distress detection models should be able to adapt to different scenarios, which is called cross-scene distress detection. Unexpected perspectives and unseen elements are potential reasons for a model's poor performance in a new scenario, resulting in higher requirements for these methods when applying existing models to new scenarios. In this case study (Li et al., 2021b), we introduce a CNN-based cross-scene transfer-learning pipeline, which combines the advantages of model transfer and data transfer.

10.4.1 Problem formulation

CNN-based cross-scene transfer-learning pipeline combines the advantages of model transfer and data transfer. In data transfer, a GAN is adopted to transfer the existing training data. In this process, some training data for the new scenes are synthesized by performing the style transfer on the original training data. In model transfer, the domain-adaptive method is used to transfer the domain knowledge generated from the existing models, which is not the scope of this case study and so is not presented. The data transfer method is proposed to synthesize images based on the existing source domain for the case of insufficient training data. The main target of this problem is to transfer pictures of the original data set to the new scene style via a generative adversarial network.

10.4.2 Model: CycleGAN-based image style transfer

In this case study, CycleGAN (Zhu et al., 2017) was selected and trained to convert one type of image into another kind of image, achieving style conversion of the original data set images to a new scene. The training set of the object detection model requires a large amount of manually annotated data, which is costly and inefficient. GAN can automatically complete this process. The two parts (the generator and the discriminator) promote each other's optimization through a game, which is a very efficient and low-cost method.

The framework of the CycleGAN is shown in Fig. 10.7. The generation network takes random samples from the feasible region as input, and its output needs to imitate the real samples in the training set as much as possible. Simultaneously, the input of the discriminating network is the real sample or the output of the generating network. Its purpose is to distinguish the output of the generating network from the real samples as much as possible. The generation network must deceive the discrimination network as much as possible. The two networks confront each other and constantly adjust their parameters. The ultimate goal is to make the discriminating network unable to judge whether the generated network's output is correct or not.

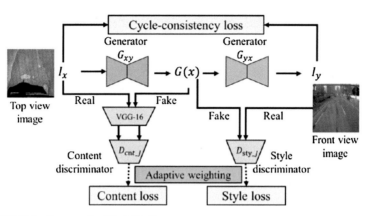

FIGURE 10.7 Framework of CycleGAN.

The most significant advantage of CycleGAN is that only unpaired data are needed during training. If a model that converts the style of an image from a sunny day to a cloudy day is trained, any sunny and cloudy image data can be used for training, instead of pairs of sunny and cloudy images at the same location. This excellent scalability is suitable for the conversion of pavement images. When training other types of GANs, pairs of distress images are needed, which means that identical distresses need to be collected in different scenarios, which is obviously impossible. The ability to train with unpaired data is the most crucial reason why we chose to use CycleGAN.

This model consists of two unidirectional GANs, as shown in Fig. 10.7. The first one contains a generator, G_{xy}, to translate image style, I_x, from domain X (labeled data set) to domain Y (unlabeled data set in a new scene) and a discriminator, D_y, to discriminate image differences. We designate the source distribution as I_X and assume we have access to a true target distribution I_Y. (In our case, samples from X and Y have consistent content yet distinct style.) Our objective is to learn a mapping $X \rightarrow Y$ (namely, the conditional $P(Y \mid X)$) to achieve style transfer. Typically, we build a generator G_{xy} instantiated by a CNN. The generator, G_{xy}, strives to make the generated image, $G(x)$, as similar as possible to I_y from domain Y. In contrast, D_y seeks to distinguish the generated sample, $G(x)$, from the real example, I_y. Adversarial losses are applied to both mapping functions as:

$$L_{GAN}\left(G_{xy}, D_y, X, Y\right) =$$
$$E\left[\log D_y(y)\right] + E\left[\log\left(1 - D_y\left(G_{xy}(x)\right)\right)\right]. \tag{10.19}$$

During training, G_{xy} tries to minimize the losses, and the discriminator, D_y, maximizes them as much as possible. The second GAN is similar to the first one, but the mapping relationship is reversed, including a generator, G_{yx}, and a discriminator, D_x. Based on two sets of loss functions, a cyclic consistency

FIGURE 10.8   Examples of the results of data transfer.

loss is added to ensure that the original image can be obtained when image I_x is converted to domain Y and then converted back to domain X. The cyclic consistency loss is shown as

$$
\begin{aligned}
L_{cyc}\left(G_{xy}, G_{yx}\right) = \\
E\left[\left\|G_{yx}\left(G_{xy}(x)\right) - x\right\|_1\right] + E\left[\left\|G_{xy}\left(G_{yx}(y)\right) - y\right\|_1\right],
\end{aligned}
\tag{10.20}
$$

where $\left\|G_{yx}\left(G_{xy}(x)\right) - x\right\|_1$ and $\left\|G_{xy}\left(G_{yx}(y)\right) - y\right\|_1$ represent the first-order distance between the two images. All the images after GAN style transfer need fast and straightforward manual screening. Unreasonable images (such as a car in the sky or a twisted road) will be filtered out and not used as training data.

10.4.3   Results

To test the proposed pipeline's performance, pavement distress data sets data set images were collected by a portable onboard camera set upon the top rear of a vehicle. This top-view data set is named TV. Because it was taken from a bird's eye perspective, most of the images' elements are road surfaces, including some tree shadows and small sections of cars. Another data set of pavement distress contains images obtained from a front view angle, FV, which were obtained with a driving recorder. The scene elements covered are more complex than in the TV data set. The shooting angle, equipment, and scene elements covered by the two data sets' images are thus different. These differences are sufficient to allow us to verify the performance of the transfer pipeline. To better test the transfer method's performance, the TV data set with simple scene elements was used as the basic data set to train the basic model. The FV data set, which contains many items that do not appear in the basic data set, was used as a new scene data set for transfer.

The effect of data transfer technology is explored based on the two datasets. Fig. 10.8 shows examples of data transfer. When evaluating the effect of Cycle-

GAN, 50 synthetic images of new scenes with 50 real images are blended. Then, these 100 images were randomly shown to three experts who were engaged in road maintenance. They were asked to make a judgment as to whether the proffered image was real or synthetic. The results show that the images synthesized by CycleGAN successfully passed the visual Turing test.

10.5 Exercises

1. Which of the following statement is wrong?
   a. When training the discriminator, we want its output to be as close to 1 as possible for real samples.
   b. When training the discriminator, we want the output to be as close to 0 as possible for fake samples.
   c. When training the generator, we want the output of the discriminator is as close to 0 as possible.
   d. When training the generator, we want the output of the discriminator to be as close to 1 as possible.
2. Which of the following statements is wrong about the latent vector z for obtaining higher quality generated samples?
   a. The need for the partition of the exerciser, the constant observation and the difference of observation.
   b. Make the latent vector z as close as possible to the distribution of the real samples, while avoiding biasing the generator towards a narrow part of the distribution of the real samples.
   c. Obtain the latent vectors's semantic features by using unsupervised learning.
3. In GAN, the discriminator network D is a binary classification network. Is it possible to change the D into a multi-class classification network, and explain your answer?
4. Write down the minimax formulation of GAN and explain it in two sentences.
5. The training process of GAN may be unstable since it may encounter the gradient vanishing and mode collapse problem. Use your own words to describe what are the gradient vanishing and mode collapse problems.
6. Try to explain the origination of the two problems from GAN's model mechanism perspective.
7. One representative application of GAN is data augmentation, especially generating very complex data. Describe five scenarios in which you can use GAN to generate transportation data.
8. What are the trade-offs between GANs and other generative models?
9. How should we evaluate GANs, and when should we use them?
10. What is the relationship between GANs and adversarial examples?

Chapter 11

Edge and parallel artificial intelligence

11.1 Edge computing concept

As a key driver for recent smart city applications, big data has been going through a radical shift—from the cloud data centers to the edge of the network, closer to where they are generated, such as smart sensors, mobile units, and Internet of Things (IoT) devices. An estimation states that about 20 ZB of data traffic can be handled by all the cloud datacenters globally in 2021, while, in total, 850 ZB of data will be generated at the network edge at the same time. If primarily relying the traditional cloud computing framework to transmit all the raw data to the cloud for processing and storage, not only will the network capacity be overwhelmed, but also privacy and cybersecurity concerns will also be escalated.

The science and engineering community is aware of the need to shift the computing workload away from the centralized cloud to the logical edge of the network. Edge computing, as a new solution to address this data explosion challenge, enables handling data generated from edge devices closer to the local clients who produce it. In edge computing, data is processed by an IoT device itself or by a local computer, rather than it all being transmitted to the cloud data centers. Edge computing is efficient in network-wide data processing because of its (1) reduced communication bandwidth, (2) reduced computation load at the centralized cloud, and (3) reduced overall cost for sensors, computation, data transmission, and maintenance. It is also more scalable, and at the same time, makes privacy protection easier by eliminating or reducing the use of raw data on the cloud.

11.2 Edge artificial intelligence

The data explosion has not only motivated the emergence of edge computing but also new advances in artificial intelligence (AI) applications. AI has empowered innovations and greatly changed the environment we live in—from online services to offline activities concerning many aspects of the world. AI models have high demands for data and computation power. The training of AI models, such as DNNs, may take days and weeks to complete. Inference with AI models is also less efficient than traditional statistical models.

Machine Learning for Transportation Research and Applications
https://doi.org/10.1016/B978-0-32-396126-4.00016-3

183

Earlier, AI models were often operated on the cloud or at least by a capable computer with powerful computation and storage supports. Within the context of edge computing, AI is also rapidly shifting from the cloud to decentralized network edges to enhance the intelligence of smart city applications. This is an unstoppable trend, but it has created new barriers we need to overcome. Given the resources and energy constraints on edge devices, it is very challenging to run AI tasks with a high computation demand. On the other hand, high efficiency is critical in many emerging applications, which poses another challenge for the design and engineering implementation of AI techniques.

Edge AI currently refers to the deployment of AI models, mostly deep neural networks, such as CNN and RNN, on the network edge. In transportation systems, the edge nodes are normally individual road users or civil infrastructure components. Vehicles' onboard units, pedestrians' wearable devices, traffic signal heads, street lights, people's smartphones, and so on, can be an edge node to perform certain intelligent transportation AI tasks. They can also be set up as edge servers to integrate computations in a local region (e.g., a road segment or an intersection). Edge AI models are trained on a separate machine (e.g., a server or a workstation) and then optimized for deployment on edge nodes for inference. It is expected that the training and inference will be integrated on the edge in the near future.

11.3 Parallel artificial intelligence

Parallel artificial intelligence is a concept associated with edge artificial intelligence, but from a system angle looking at the subject. Since the tasks are distributed from a centralized server to the network edge nodes, it is natural that the computation shifts from a sequential manner to a more parallel manner. Tasks are broken down into pieces. Parallel AI is multi-tasking that can occur either on an individual device or multiple devices. On a single-edge device, e.g., an IoT computer, multi-thread system architecture can be designed to operate a variety of tasks. For example, road mask segmentation and traffic flow detection can be two parallel tasks operating on the same edge roadside unit in different threads. While multi-threading puts more computation resource restrictions on every single task, communication and information sharing between different tasks are easier. Multiple-device parallel AI aims at large-scale applications. Proper system architecture with a computation and communication mechanism for subsystems is necessary. Subsystem cooperation and data sharing among multiple entities require careful design to make the entire system work in an efficient and effective manner. National ITS architecture and regional ITS architectures currently specify the information flow and roles of participants for a specific ITS task. It would be helpful in the future edition of ITS architectures to specify more detailed agency cooperation towards parallel intelligence in ITS applications. Lately, federated learning, as a distributed and parallel learning mechanism, has been emerging in the ITS field to support parallel and edge AI, with an emphasis on data privacy protection.

11.4 Federated learning concept

Federated learning is a method for distributed model training across many devices. A single machine-learning model is stored in the central cloud server. Each device has its own data, and they do not share raw data with one another due to barriers, such as privacy issues, regulations, or technology constraints. The data is stored locally and is often heterogeneous as well. Federated learning, though conducts learning tasks in a distributed manner, is different from traditional distributed learning in a few ways. Traditional distributed learning happens on the cloud for big data processing. The vast amount of data is usually the same dataset yet distributed on multiple machines to process the data in a parallel manner. However, these data are expected with the same distribution (e.g., the same mean and variance). In federated learning, data are from multiple worker nodes. Some worker nodes can be unstable, and the data stored across all worker nodes are imbalanced and not independent and identically distributed (IID). At the same time, users have control of their devices (e.g., cell phones) and can terminate data sharing at any time.

Assume there are multiple edge nodes $node_1$, $node_2$, ... $node_n$. $Node_i$ has the private dataset D_i. The loss function of the machine learning model on the central server is $L(\cdot)$. The devices train local models using local datasets and synchronize metadata on the central cloud server, instead of transmitting raw data to the cloud. Typical metadata transmitted to the cloud server is the local weight w_t^i at time t for $node_i$. In one synchronization, node computes its updated weight w_{t+1}^i with step size γ_t as follows:

$$w_{t+1}^i = w_t - \gamma_t \cdot \frac{\partial f(w_{t,i})}{\partial w} \quad (i = 1, 2, \cdots, n). \tag{11.1}$$

Note that this local update can run one or multiple iterations before the metadata transmission. On the cloud, it receives the weights from all nodes and uses an aggregated function $A(\cdot)$ for the integration of all the uploaded weights. In this way, the weights for the overall model are updated for the new round:

$$w_{t+1} = A\left(w_t^1, w_t^2, \cdots, w_t^n\right). \tag{11.2}$$

A simple and effective aggregation method is federated averaging, which basically takes the average of all the local weights at the cloud server (McMahan et al., 2017).

11.5 Federated learning methods

11.5.1 Horizontal federated learning

Horizontal federated learning (HFL) is introduced in which datasets share the same set of features yet different samples. For example, two cities have different roadways and sensor groups, but the data needed for traffic management is similar and the sensor features are the same (e.g., volume, occupancy, and speed). In horizontal federated learning, agents collaborate and share subsets of updates of parameters. Agents update the parameters locally and upload them to the cloud, thus jointly training the centralized model with other agents.

HFL is confronted with several unique challenges when compared to other machine learning models. First, in Non-Independent Identical Distribution (Non-IID), the distribution of data in each device is not representative of the global data distribution. The category of samples on each agent could vary. Dealing with Non-IID data may significantly increase the processing complexity and error rates. Second, the data distribution among different agents may be unbalanced. The number of samples could vary because some agents produce much more data than others. The number of samples for each category could also be different: Two agents both have category A and category B, but the ratio of sample numbers A over B on agent 1 is far larger than that on agent 2. Unbalanced data creates problems of heterogeneity and increased training complexity. HFL can be denoted as

$$X_i = X_j, \quad Y_i = Y_j, \quad I_i \neq I_j, \quad \forall D_i, D_j, \ i \neq j.$$

11.5.2 Vertical federated learning

In HFL, data have various sample spaces yet share same feature space. Vertical federated learning (VFL) is another FL method in which the data have the same sample space but a different feature space. The data among agents are vertically partitioned. Data samples are shared, but features are distributed on multiple devices. For example, if a car insurance company would like to better estimate customers' discount rates, they tend to integrate data from outside their insurance company to train a model, e.g., from banks and hospitals. However, these organizations cannot share the raw data with the insurance company due to regulations on privacy and security. In order to share information, separate models should be trained at these different organizations, and only the intermediate outputs and their gradients will be shared. This is also different from HFL, in which the exchanged messages are global and local model parameters. Note that this is a typical VFL example because every organization has all the sample IDs (insurance policyholders) yet the data features are handled separately.

VFL has the advantages of being flexible and keeping local secrets. VFL can explore deeper feature dimensions and acquire a better learning model. Up to now, while there has been much less effort in investigating challenges and solutions in VFL than in HFL, the research gap and motivation have inspired some researchers to put efforts into VFL. Wei et al. recently proposed a general VFL process (Wei et al., 2022), including seven steps:

- Private set intersection;
- Bottom model forward propagation;

- Forward output transmission;
- Top model forward propagation;
- Top model backward propagation;
- Backward output transmission;
- Bottom model backward propagation.

11.6 Case study 1: parallel and edge AI in multi-task traffic surveillance

11.6.1 Motivations

This case study is treated in our paper Ke et al. (2022). Smart transportation surveillance in city-scale applications has a high demand for computing services to process, analyze, and store big transportation data. Conventionally, data were mostly born and processed in cloud datacenters. Cloud computing was widely recognized as the best computing service for big data processing and AI tasks. The network bandwidth and capacity could be overwhelmed in modern smart transportation applications if one employs the traditional cloud computing scheme; also, privacy and security issues are further raised.

Research on developing edge computing systems for smart transportation is still in an early stage. The major challenge is the sharp contrast between the demand for high efficiency/intelligence and the computing/communication constraint on IoT devices, even for a single task. The state-of-the-art solutions are model compression and cloud-edge computing integration. Model compression reduces the size of the neural networks or other models via pruning or quantization, therefore, increasing the inference speed. Cloud-edge computing integration splits the computation, thus balancing the workload on the edge and the cloud to achieve optimal performance.

As a critical component of the smart city, smart transportation is facing opportunities and challenges in the new era of AI and advanced computing matching the trend of being more efficient and integrated. Current edge ITS systems mainly focus on a single task without the ability to carry out multiple surveillance tasks in a real-time and parallel manner. To fill this gap, this study explores a multi-task system for road and traffic surveillance with parallel edge artificial intelligence. Specifically, this system makes contributions to the field with a multi-thread edge architecture, an efficient clustering method for road mask extraction, an AI trigger method aiming at parallel edge processing, a vehicle detector, and a road surface-condition classifier.

11.6.2 Parallel edge computing system architecture

Unlike some state-of-the-art architecture for splitting the computation loads to the edge and cloud, the proposed architecture (see Fig. 11.1) is designed to operate mostly on the edge device, and, in this case, the Raspberry Pi 4 edge computer. There are three parallel threads operating independently in this

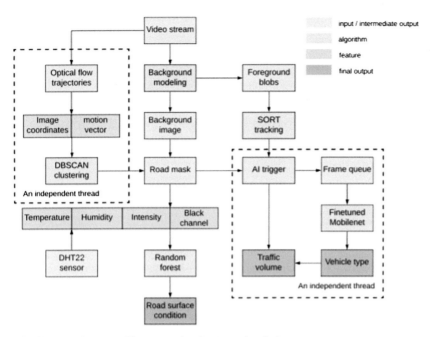

FIGURE 11.1 System architecture on the edge computing device.

system: the main thread, the road mask extraction thread, and the traffic flow-detection thread. The reason for designing three threads instead of one is that the road mask-detection task and the traffic flow-detection task have high computation complexity. This architecture design, coupled with the road mask extraction method and the AI trigger method, can address this issue, thus enabling real-time processing on the edge device.

In the main thread, background modeling and SORT tracking are implemented to generate foreground image and blobs for detecting and tracking moving objects. At the same time, the background image (without moving objects) is used for roadway surface- condition classification (dry, rainy, snowy). The road mask received from the road mask- extraction thread serves as a filter to isolate the road pixels in the background image. Thus, moving traffic on the road and non-road regions in the camera view are filtered out in the road surface condition-classification task for improved accuracy. Four features, including two image features (intensity and black channel) and two environmental features (temperature and humidity) are used by a random forest model for the road condition classification.

The road mask-extraction thread also takes the raw video stream as input. The intuition behind this new road mask-extraction method is that human eyes can tell where the road regions are at night, with the help of vehicle trajectories. Optical flow tracker extracts moving pixels at night in traffic videos with good accuracy. In this case study, the optical flow tracker is applied to extract moving pixels to roughly get the vehicle trajectories, and then to cluster the pixels into road regions for various traffic directions with location and motion features and DBSCAN clustering. DBSCAN clustering is chosen since it determines the number of clusters and filters out outliers.

The third thread is for traffic flow counting and classification. We designed an AI trigger method to select, store, and process just a small portion of the original video frames, which are the video frames of interest. Specifically, when moving objects are defined and tracked in predefined regions on the road, the frame will be stored in a queue waiting to be processed. The Mobilenet fine-tuned on the MIO-TCD traffic surveillance dataset is implemented to locate and classify road users in the predefined road region. With the tracked and classified objects, traffic volume data for each type of road user are obtained.

11.6.3 Algorithms and results

In the parallel architecture, the road mask-extraction algorithm serves two purposes: to help locate the road regions so that the image features (intensity and black channel) for the road surface condition classification are more representative, and to identify the number of travel directions and roughly locate the regions for each direction to assist traffic volume detection. We adopt the clustering of motion vectors extracted from the optical flow on the moving traffic, instead of using the appearance features of the road itself. The intuition behind this method is an approximation of human eyes identifying road regions at night by watching the moving traffic. The DBSCAN clustering algorithm is applied to the filtered optical flow motion-vectors to extract the road regions and classify the travel directions using the positions, moving direction, and speed of the motion-vectors. DBSCAN is also able to filter out outliers that do not belong to any of the clusters, thereby auto-removing false positives. Closing operation in image morphology is applied to fill the "holes" within the clusters. Two example results are shown in Fig. 11.2.

In the road surface condition-classification algorithm, we selected four scalar features: temperature, humidity, median image intensity, and median image black-channel value. The intensity value is selected because it can differentiate snowy conditions and non-snowy conditions since the intensity values are high in snowy conditions where the color is white in many places. The dark channel value is defined as the minimum value among R, G, and B channels for a given pixel. For most objects in nature, the dark channel value is pretty small, otherwise, the object will be white. This is also the case for most road surfaces. However, for rainy or snowy conditions where there are many reflections or white pixels, the dark channel can be relatively higher. We tested multiple traditional machine learning classifiers on the features and selected the random forest in the end. Experimental data is collected from the WSDOT Traveler API (see Figs. 11.3 and 11.4).

FIGURE 11.2 Motion-vector clustering method to extract road mask and identify traffic directions.

FIGURE 11.3 WSDOT camera distribution (left), WSDOT weather station distribution (middle), and the selected locations for testing the road surface classifiers (right).

Regarding the AI-trigger-based traffic flow monitoring, the training of the Mobilenet classifier is standard. Transfer learning is applied with the MIO–TCD surveillance dataset to enhance the classifier. We combine the original class labels into four road- user categories (car, truck, bus, and cyclist) and the background. As shown in the design system architecture, this Mobilenet classifier operates in a separate thread with a queue storing those "triggered frames" for further classification and validation. As shown in Fig. 11.5, when moving blobs are detected and tracked in the main thread, they are first filtered using the road mask to determine if they are in the ROI. If yes, the current frame will be pushed to the queue, and the AI thread will be triggered to further validate if the

StationID":3274 ReadingTime":"\/Date(1592798838000-0700)\/"   TemperatureInFahrenheit":74.12
RelativeHumidity":26   "BarometricPressure":null   WindSpeedInMPH":12
StationID":3274 ReadingTime":"\/Date(1592799737000-0700)\/"   TemperatureInFahrenheit":73.58
RelativeHumidity":25   "BarometricPressure":null   WindSpeedInMPH":11
StationID":3274 ReadingTime":"\/Date(1592800638000-0700)\/"   TemperatureInFahrenheit":72.86
RelativeHumidity":27   "BarometricPressure":null   WindSpeedInMPH":15
StationID":3274 ReadingTime":"\/Date(1592800638000-0700)\/"   TemperatureInFahrenheit":72.86
RelativeHumidity":27   "BarometricPressure":null   WindSpeedInMPH":15
StationID":3274 ReadingTime":"\/Date(1592801537000-0700)\/"   TemperatureInFahrenheit":71.96
RelativeHumidity":30   "BarometricPressure":null   WindSpeedInMPH":17
StationID":3274 ReadingTime":"\/Date(1592802438000-0700)\/"   TemperatureInFahrenheit":71.60
RelativeHumidity":30   "BarometricPressure":null   WindSpeedInMPH":15
StationID":3274 ReadingTime":"\/Date(1592802438000-0700)\/"   TemperatureInFahrenheit":71.60
RelativeHumidity":30   "BarometricPressure":null   WindSpeedInMPH":15

FIGURE 11.4 Surveillance camera image samples for rainy, snowy, and dry road surfaces, and weather data sample file collected from the WSDOT API for Station 3274.

FIGURE 11.5 AI trigger method that uses traditional detection and tracking as the trigger, and the finetuned Mobilenet as the final classifier for traffic flow detection.

moving objects are road users or background (moving background blobs may be caused by camera shaking or sudden lighting changes) and then count the road users of each type. Since we track the road users, a new object ID will add one count to a specific road-user type. The system architecture that separates the AI classifier into a separate queue enabling capturing the video temporal information in real-time and automatically finalizes the classification and volume counting a few seconds later in dense traffic-volume conditions (i.e., when the queue is not empty).

11.7 Case study 2: edge AI in vehicle near-crash detection

11.7.1 Motivations

This case study is from our work reported in Ke et al. (2020a). Herbert Heinrich proposed the relationship among major injury, minor injury, and no injury incidents (1 major injury incident to 29 minor injury incidents to 300 no injury incidents) (Heinrich et al., 1941). The linear relationship holds for traffic safety events as well, though the exact ratio may be different (Klauer et al., 2006; Talebpour et al., 2014; Makizako et al., 2018). The near-crash scenario has two major properties that make it valuable for a variety of research and engineering topics: (1) It reflects the underlying causes of the incidents that resulting no or minor losses; (2) It occurs on many more occasions than the accidents involving contact.

Near-crash data is irreplaceable in smart transportation applications. In traffic safety research, near-crash data is the surrogate safety data for studying and assessing the safety performance of particular locations or scenarios This is because a certain amount of data is required to feed either traditional statistical analytical methods or emerging machine learning models. For example, to understand the safety-related designs at a roadway intersection, the collision data might be far from sufficient to support models to reach any statistically significant conclusions. Near-crash data fills this void with the aforementioned two properties.

With the emergence of concepts and technologies in the intelligent vehicle (IV) and autonomous vehicle (AV), near-crash events become an even more valuable data source for not only traditional traffic safety research but also IV and AV safety. The latest AVs have been demonstrated to be able to handle most situations they may encounter. However, the lack of corner cases for training and testing is a major bottleneck that is slowing down the pace to achieve the goal of Level-5 (L5) fully autonomous driving. Corner cases belong to subsets of near-crashes; they rarely occur, such as pedestrians walking across a highway, but can cause severe losses. Leading research in the AV field is focused on speeding up the generation of corner cases in simulation by leveraging historical crash or near-crash data for model calibration and training.

Three key challenges remain for near-crash detection and data collection. Firstly, near-crashes are still rare events that require intensive computing, transmission bandwidth, and data storage services on the original large data sources. Secondly, existing methods documented in papers and reports rely on manual checking or post-analysis of the original data sources, which are inefficient. Thirdly, while the state-of-the-practice commercial collision avoidance systems (e.g., the aforementioned Shield + system) can serve for the purpose of near-crash data collection, their purchase and maintenance costs are very high, and each product lacks transferability and scalability due to the design for a certain type of vehicles or roadside units.

Under the context of edge computing, real-time video analytics is regarded as the killer app, given the restricted computing resources on edge devices and the property of video data requiring large capacity due to their enormous 3D matrices. To this end, this case study introduces a light-weight edge computing system for real-time near-crash detection and data transmission with normal dashcams and network bandwidth. The system is a low-cost and stand-alone system that is backward-compatible with existing vehicles. It is being developed based on the Nvidia Jetson TX2 Internet-of-Things (IoT) platform.

11.7.2   Relative motion patterns in camera views for near-crashes

Relative motions between the ego vehicle and other road users are important cues for near-crash detection using a single camera. Relative motion patterns, as well as the relationship between a pattern in the camera view and its corresponding pattern in the real world, must be understood (see Fig. 11.6). The relative motion patterns between two road users vary from case to case. Roadway geometry, road user's behavior, relative position, traffic scenario, etc. are all factors that may affect the relative motion patterns. For example, from the ego vehicle's perspective, its motion relative to a vehicle it is overtaking in the neighboring lane and that to another vehicle it is following in the same lane are different.

Relative motion that has the potential to develop into a crash/near-crash is characterized from the ego vehicle's perspective as the target road user moving towards it. This kind of relative motion is shown as a motion vector of the target road user moving vertically towards the bottom side of the camera view. Examples are shown as solid red arrows in Fig. 11.6. In the real-world top view, the three solid red arrows represent the relative motions between the ego vehicle and each of the three road users (a pick-up truck, a car, and a pedestrian). Each of the three camera sight lines aligns with a relative motion vector (Z2, Z4, and Z7). In the camera view, the lines of sight are shown as vertical bands. The relative motion vectors for near-crashes in the top view correspond to vectors moving towards the bottom in the camera view aligning with Z2, Z4, and Z7.

Two road users have an apparent relative motion at any time. In addition to the near-crash cases just defined, other patterns may occur. First, a target road user may move towards the ego vehicle, move away from the ego vehicle, or stay at the same distance to the ego vehicle. These can be identified as object image-size changes in the camera. This property will be utilized later in our approach. An image size decrease or no size change would not indicate a potential crash or near-crash. For size increases, there are three cases. The first cases constitute the potential crashes, shown as the solid red arrows in Fig. 11.6. The second are the warning cases, shown as the dashed orange arrows, in which the relative motion is towards the center line of sight of the camera (the pick-up truck and the pedestrian), or the relative motion is slightly different from the solid red arrow, while the target road user is at the center line of sight (the car). The warning cases could develop into crashes if there are slight changes in the speeds or

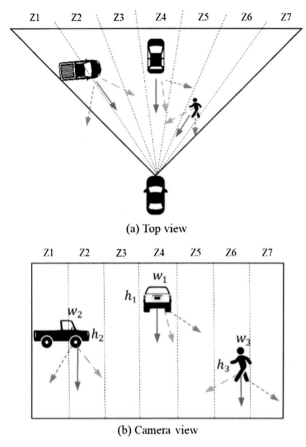

(a) Top view

(b) Camera view

FIGURE 11.6 Relative motion patterns in a dashcam view.

headings of either the target or the ego vehicle. The third case is the safety case that relative motion is moving away from the center line of sight, shown as the dotted green arrows in Fig. 11.6.

11.7.3 Edge computing system architecture

The overall system architecture on the edge computing platform is shown in Fig. 11.7. The two major functions of the system are near-crash detection and data collection. Given the real-time operation requirement, the design is concise enough to be highly efficient and sophisticated enough for high accuracy and reliability. The near-crash detection method also should be insensitive to camera parameters to accommodate large-scale deployments.

The system is implemented in a multi-thread manner. Four threads are operating simultaneously: the main thread, data transmission thread, video frame

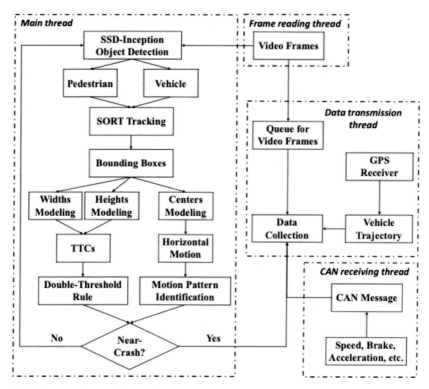

FIGURE 11.7   Edge architecture for near-crash detection on intelligent vehicles.

reading thread, and CAN receiving thread (CAN = Controller Area Network). The proposed near-crash detection method is implemented in the main thread. When near-crash events are detected, a trigger will be sent to the data transmission thread, and it will record video frames from a queue (a global variable) and other data that are associated with the near-crash event. The third thread for video frame reading stores the latest video frame captured from the camera in another queue and will dump previous frames when the capturing speed is faster than the main thread's frame-processing speed. The CAN receiving thread provides additional information for each near-crash event with the ego vehicle's speed, brake, acceleration, etc.

The proposed architecture ensures that the system delay is small. The frame reading thread ensures that the main thread reads the latest frame captured by the camera by not accumulating frames. The data transmission thread is designed as an individual thread to handle data transmission so that the main thread operation is not affected by the network bandwidth. The CAN receiving thread is for additional information collection, and the purpose for separating it as another individual thread is the consideration of system function extension. The proposed system can communicate with other systems via this thread while not affecting the performance of itself.

11.7.4 Camera-parameter-free near-crash detection algorithm

The main thread starts with applying a deep learning-based object detector to every video frame. The object detection creates bounding boxes and identifies the types of road users in each video frame. To associate the information from each frame and find each road user's movement, the standard step following object detection is object tracking, in this case, SORT tracker. An object appears larger in the camera view it approaches the camera, and smaller as distance to the camera increases. In this study, the proposed approach for TTC estimation mainly considers: (1) leveraging the power of recent achievements in deep learning, (2) making the computation as efficient as possible to support real-time processing on the Jetson, and (3) transferability to any dashboard camera without knowing the camera's intrinsic parameters.

Given a video with a frame rate of 24 FPS, the next frame is captured in less than 0.05 seconds. Thus, for size change detection, we use more frames to compensate for the noise in each frame and increase the time interval for the detection. Linear regression is used for bounding boxes' heights or widths over a group of consecutive frames. We found that 10–15 frames are enough to compensate for noise and the time associated with 10–15 frames is still small enough (about 0.5 second) to assume that the road user's motion is consistent. Therefore, the input to the linear regression is a list of heights or widths extracted from the bounding boxes, and the slope outputted by the regression will be the size change rate. With a similar mathematical derivation shown in Chap. 7 of Ke (2020), the TTC value can be estimated without camera parameters using the following equation, where D_t, V_t, s_t, r_t are the distance between the ego vehicle and a target road user, the relative longitudinal speed, the size of the bounding box, and the size change rate of the bounding box, respectively, all at time t:

$$TTC = \frac{-D_t}{V_t} = \frac{s_t}{r_t}. \qquad (11.3)$$

According to Eq. (11.3), TTC can be calculated as the size of the bounding box at time t divided by the size change rate at time t, and it is not related to the focal length or other intrinsic camera parameters. The TTC value can be either positive or negative, where being positive means the target is approaching the ego vehicle, and being negative means it is moving away from the ego vehicle.

11.7.5 Height or width

There are two options for the size of the road user in the camera view: height and width. We argue that height is a better indicator than width. From the ego vehicle's perspective, one might observe a target vehicle's rear view, front view, side view, or a combination of them, depending on the angle between the two vehicles. That is to say, the bounding box's width change may be caused by either the relative distance change or the view angle change. For example, when the ego vehicle is overtaking the target vehicle, or the target vehicle is making a turn, the view angle changes and will lead to the bounding box's width change.

However, the bounding box's height of the target vehicle is not influenced by the view angle; it is solely determined by the relative distance between the two vehicles. Similarly, a pedestrian walking or standing on the street may have different bounding box widths due to not only the relative distance to the ego vehicle but also the pose of the pedestrian; but the height of a pedestrian is relatively constant.

Despite the challenge of using width to determine an accurate TTC, it still provides valuable information. Since we are using only less than one second of frames for the calculation, the view change does not contribute as much as the distance change, so width still roughly indicate the longitudinal movement of the road user. This is very important in some cases. For instance, a vehicle moving in the opposite direction of the ego vehicle is truncated by the video frame boundary. In this case, the height of the vehicle increases, while the width decreases. This is not a near-crash case at all, but the TTC can be very small and falsely indicate a near-crash by only looking at the height change.

We propose a double-threshold rule: If the TTC threshold for determining a near-crash is δ, we will set δ as the TTC threshold associated with the height regression. At the same time, we have another TTC threshold ϕ associated with the width regression. The second threshold ϕ is to ensure that the width and height changes are in the same direction. The rule is represented as in Eq. (11.4):

$$0 < \frac{h}{r_h} < \delta; \quad 0 < \frac{w}{r_w} < \phi; \quad \delta < \phi, \tag{11.4}$$

where r_h and r_w are the change rates for height h and width w.

11.7.6 Modeling bounding box centers for horizontal motion pattern identification

As shown in Fig. 11.6, there are three scenarios for the case that a road user approaches the ego vehicle; they correspond to potential crashes, warnings, and safe scenarios. Besides TTC, these scenarios can be differentiated with the relative horizontal motion between the ego vehicle and the target. This also needs to be calculated with computationally fast methods. We propose to apply another linear regression using a list of bounding box's centers of the target road user. The regression result would be able to indicate the moving direction of the road user in the camera view.

In general, when the target's location is closer to the bottom and closer to the center line of sight, the risk of a collision is higher, so the threshold for the moving direction ω is looser. We propose a rule to show this judgment as in Eq. (11.5):

$$\alpha < \omega \times (C_x - C_{los}) \times (B_y - B) < \beta, \tag{11.5}$$

where C_x is the center's x coordinate, C_{los} is the center line of sight, B_y is the bottom side of the bounding box, and B is the bottom of the video frame. Since cameras have different resolutions, $C_x - C_{los}$ is normalized to $[-1, 1]$, and $B_y - B$ is normalized to $[0, 1]$. The two thresholds are α, β. α should be set to negative to capture the potential warning scenarios (the orange dashed arrows in Fig. 11.6). Also, β should be just slightly larger than zero to capture the potential crashes (the solid red arrows in Fig. 11.6) and filter out most of the safe scenarios (the green dotted arrows in Fig. 11.6). Eqs. (11.4) and (11.5) together identify near-crash events.

11.7.7 Experimental results

The system consists of an Nvidia Jetson TX2, a dashcam (can be USB camera or IP camera), a GPS receiver, an in-vehicle power inverter, a PEAK CAN adapter for CAN bus communication, an external circuit based on Arduino board for auto bootup, a shell for the Jetson device, an ethernet cable, two power cables, an internet switch, mounting materials, and a cloud server. The Nvidia Jetson device is the key processing unit of the system, running the near-crash detection, video streaming, data transmission, data fusion threads, and algorithms. The Jetson was powered by in-vehicle (either car or bus) 12V-DC power through the power inverter. The Arduino circuit is connected to the Jetson, and, when the vehicle's power is on, it will auto boot up the system. Fig. 11.8 shows the system and testing buses for the real-world test. From top to bottom: the systems ready to be installed (before installation), three of the testing buses at Pierce Transit, the radio box behind bus driver's seat where the system works, and the system being tested in the radio box.

Essentially, near-crash is a type of traffic anomaly. To evaluate the proposed method's accuracy, we used the evaluation process of the traffic anomaly detection task (Track 4) of the 2021 AI City Challenge as the reference. First, the task dataset has 100 video clips with some anomalies. It is unknown exactly how many anomalies are in the test dataset, but the number is between 0 and 100, as mentioned in the introduction to Track 4. Likewise, we made a local test dataset with 5,000 video clips with 500 near-crash events. As aforementioned, the test videos were from online resources and dashboard cameras on private cars and transit buses. This data set is not being published due to potential privacy and copyright issues. There is a plan to create such a video data set for near-crash detection in the future.

We manually labeled all the near-crash events with their occurrence videos and times. As in AI City Challenge Track 4, we defined a true-positive (TP) as a predicted near-crash within 10 sec of the true near-crash. A false-positive (FP)

FIGURE 11.8 The system prototypes, buses for the real-world testing, and the bus radio box where the system works.

is a predicted near-crash that is not a TP for a near-crash. A false-negative (FN) was a true near-crash that was not predicted. We used the F1 score to evaluate accuracy. F1 score was the harmonic mean of the precision and recall, where the best value $= 1$ and the worst value $= 0$:

$$F1 = 2 \times \frac{precision \times recall}{precision + recall} = \frac{2TP}{2TP + FP + FN}. \tag{11.6}$$

FIGURE 11.9 Sample near-crash detection results, where red bounding boxes (mid gray in print version) indicate the potential conflict with the road user. Each row is a four-frame sequence of one near-crash event.

Sample near-crash detection results are shown in Fig. 11.9. The top three rows were three vehicle–vehicle near-crashes, and the bottom two rows were two of the vehicle–pedestrian near-crashes. The bounding boxes turned red to indicate a predicted near-crash, while other detected road users had green bounding boxes. A few more sample detection results can be found in the video published at https://www.youtube.com/watch?v=9NGo4Ef59i0.

FIGURE 11.10 Sample CAN data collected from a vehicle–vehicle near-crash (left) and a vehicle–pedestrian near-crash (right) on May 7, 2021.

Our system correctly predicted 496 out of the 500 labeled near-crashes and missed just 4. It generated 18 FPs in the 5,000 video clips. Based on Eq. (11.6), the final F1 score was 0.988, and the average processing speed with Max-N mode was about 18 frames-per-sec (FPS). The performance was promising, considering that we intentionally included a variety of near-crash scenarios and some very challenging cases in the data set. There were adverse weather conditions (e.g., foggy, rainy, snowy), nighttime situations, traffic congestion, urban/rural traffic scenes, and so on. It is worth mentioning that the 5,000 video clips are from various cameras, and the proposed system knew nothing about the camera parameters of any of these cameras. This result benefited from the near-crash detection method. It again highlighted the possibility for low-cost and highly efficient large-scale application of the edge computing system to partially fulfill the purposes of safety data generation, IV corner case collection, and collision avoidance.

We carefully examined the FN and FP cases and summarized the causes. One of the four FNs that the system missed was a vehicle–pedestrian near-crash at night on a rural freeway with no streetlight. The pedestrian violated traffic rules by crossing the freeway, and the driver did not see him until he had almost run into him. The pedestrian was entirely in the dark so the object detector missed him. Though there were more FPs than FNs, we considered only 18 FPs out of 5,000 video clips acceptable and encouraging, given the tradeoff in the efficiency of the system. While the proposed near-crash detection method can compensate for bounding box size noise in most cases, it was not perfect. In the fourth case (the fourth row) of Fig. 11.9, right before the correct detection of this vehicle–pedestrian near-crash, there was a vehicle-vehicle FP caused by a significant error in vehicle size detection. It was included in our YouTube demo video. To further improve detection performance, a practical solution is to enhance the algorithm by further incorporating CAN messages into the detection algorithm. Sample CAN data, including bus speed, deceleration (can be calculated from speed), and brake switch associated with two sample events on May 7, 2021, are shown in Fig. 11.10. Throttle percentage data were collected as well, but not shown in the figure because they were zero in both events.

11.8 Case study 3: federated learning for vehicle trajectory prediction

11.8.1 Motivation

Data privacy has been given increasing attention in smart city applications with the rapidly improved ability in big data analytics. McMahan et al. (2017) first proposed the Federated Learning (FL) technique, which later developed into a mainstream technology of shared learning with privacy.

At present, FL has been used in many fields, such as mobile keyboard prediction (Hard et al., 2018), Healthcare (Xu et al., 2021), etc. Feng et al. (2020) proposed PMF and Li et al. (2020a) proposed STSAN using FL to predict human mobility. The most widely used federated learning algorithm is the FedAvg algorithm (McMahan et al., 2017). Li et al. (2019a) proposed the problem of Q-FFL, targeting a fairer algorithm. A semi-supervised federated learning (SSFL) framework is proposed to accurately identify travel patterns without using the user's original trajectory data or relying on significant data tags.

FL is explored in this case study for vehicle trajectory data mining. Since vehicle trajectory data are collected mostly from private vehicles, user privacy should be considered while making the data useful for traffic studies and operations. Data similar in locations or start times can have inner links; therefore, besides using FL for different users or clients, the case study proposes a new FL method called STFL, which is an FL model with a second layer to consider spatial and temporal features. We design and integrate two FL algorithms, i.e., space trajectory FL (s-FedWvg) and time trajectory FL (t-FedWvg). The model balances privacy protection and spatiotemporal feature sharing and learning.

In summary, in this case study, FL is applied to vehicle trajectory data, and the transformer model is combined to predict the trajectory location. It enables users to train the model together without sharing private information, and get better results compared to training locally. Considering the spatial characteristics of trajectories, FL based on spatial division(s-FedWvg) is carried out inside each client to extract the internal characteristics of trajectories in multiple regions to improve the learning effect. In addition, another algorithm of FL, namely FedWvg, is adopted. Compared with the FedAvg[1] algorithm, our model allocates different weights according to the different trajectory numbers. Time-division-based FL(t-FedWvg) is carried out inside the client to extract the track within the same time period for joint learning to improve the learning effect. FedWvg is also used, and s-FedWvg and t-FedWvg are combined to become the STFL.

11.8.2 Methodology

The proposed STFL model is divided into two layers (see Fig. 11.11):

- In the first layer, each client is a different user that needs data privacy protection. During each communication round, each client trains the model locally,

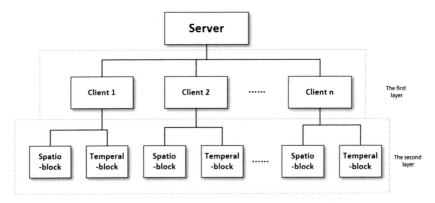

FIGURE 11.11 The general structure of the model.

then uploads the parameters to the central server, which then weights and averages the parameters and returns the values to each client.

- In the second layer, we designed time federated learning block and space federated learning block for each client. We believe that the design of these two parts can capture more of the temporal and spatial characteristics of the trajectory. In the spatio-part, data in each client is allocated into some small *second* clients based on the location of the trajectory. In the temporal part, data in each client is allocated into some small *second* clients based on the time of the trajectory.

The clients of the first layer are different users who need data privacy protection. They train the model together through federated learning.

In this first layer, let $C = \{C_1, C_2, \cdots, C_n\}$ represents n different clients, $D_i = \{D_{i1}, D_{i2}, \cdots, D_{ij}\}$ represents j pieces of data in the ith client, and w is the global parameter during training. T is the number of communication round. The training process is as follows:

- **Initialization:** In each training round, the server randomly selects n clients for interaction in this round. The server sends the current parameters w to each participating client to update its parameters.
- **Training:** The m selected client performs training locally. After the training is complete, parameter w for each client is uploaded to the central client.
- **Aggregation:** The central server uses FedAvg to assign parameters for the next round of training. The formula is as follows:

$$w = \frac{\sum_{i=1}^{m} w_i}{m}. \tag{11.7}$$

After the calculation, the parameters are reserved for allocation to participating clients in the next round.

Algorithm 1: FedAvg in trajectory.

1: Update the global parameter w to each client
2: **for** each round $t \in [1, T]$ **do**
3:     randomly choose $C_1, C_2, \cdots, C_m$ from C
4:     **for** each client C_x **do**
5:        ClientTrain(w, D_x);
6:     **end for**
7:     $w_t = \frac{\sum_{i=1}^{m} w_i}{m}$
8: **end for**

Algorithm 2: Spatio-Federated learning.

Spatio-Clustering Input:
    Data set $D = \{D_1, D_2, \cdots, D_n\}$
    x, the number of groups to be divided
Spatio-Clustering Output:
    Result x client groups $G = \{G_1, G_2, \cdots, G_x\}$
    i^{th} group with j clients $G_i = \{D_{i1}, D_{i2}, \cdots, D_{ij}\}$
1: Update the global parameter w to each client group
2: **for** each round $t \in [1, T]$ **do**
3:     randomly choose $G_1, G_2, \cdots, G_m$ from G
4:     **for** each client group G_x **do**
5:        ClientTrain(w, D_x);
6:     **end for**
7:     i^{th} weight for i^{th} client group $\alpha_i = \frac{n_i}{n_{total}}$
8:     $w_t = \frac{\sum_{i=1}^{m} \alpha_i w_i}{m}$
9: **end for**

We designed a spatio-block to extract more spatial features of the trajectory. Based on the spatial characteristics of the trajectory, K-means is used to divide the regions of trajectories. This helps the model learn better spatial properties. We take the location of the starting point of each trajectory as the basis for the division of the K-means algorithm because we believe that the trajectory of the same starting point may have similar characteristics. At the same time, we believe that FedAvg algorithm is not enough to achieve a better effect on the model, so we use a weighted average algorithm(FedWvg) to determine the proportion of the client's contribution according to the amount of data.

- **Client Division:** For each trajectory data $D = \{d_1, d_2, \cdots\}$, we select the location $location_1$ of the starting point d_1 as the dividing basis, and use k-means algorithm to divide the trajectory to form different clients.

- **Training:** m clients are randomly selected. The server distributes the parameters of this round to the clients. After the parameters of the clients are updated, each client starts the training of this round.
- **Aggregation(FedWvg):** After all clients are trained, parameters are updated to the server. Set the number of datas of m clients involved in training as $n_1, n_2, \cdots, n_m$, its parameters are $w_1, w_2, \cdots, w_m$, and we set the weight of ith client group:

$$\alpha_i = \frac{n_i}{n_{total}}. \tag{11.8}$$

We believe that the total number of tracks m owned by each client is different, so FedWvg is used:

$$w_t = \frac{\sum_{i=1}^{m} \alpha_i w_i}{m}. \tag{11.9}$$

Based on the time characteristic of the trajectory, we designed a temporal-block. We divide the data according to the time of each trajectory. This helps the model learn better time characteristics. For example, a Monday trajectory might have certain characteristics that are different from a Saturday trajectory. Also, the distribution of trajectory during the day will be different from that at night. Therefore, taking these two points into consideration, we distinguish the trajectories from Monday to Sunday, and divide each day into x different clients in several hours according to the degree of time aggregation of the tracks. Finally, the weighted average algorithm(FedWvg) is also used to determine the weight of each client participating in updating parameters.

- **Client Division:** For each trajectory $D = \{d_1, d_2, \cdots\}$, we choose t_1 at the starting point d_1 as the dividing basis, and first divide it into seven different groups according to the day of the week. In each group, the clustering algorithm K-means is used to divide it into x group, so $7x$ clients are finally obtained.
- **Training:** The same in the spatio-block.
- **Aggregation(FedWvg):** The same in the spatio-block.

In general, multiple physical features can make model training more effective. However, the normal trajectory data set usually only has several features $d = \{id_{traj}, id_{user}, x, y, t\}$, id of trajectory, id of user, longitude, latitude, and time. Therefore, we extend several physical features according to the original data, and can choose them according to the actual situation in the training (see Table 11.1).

- **number:** According to the range of all coordinate points, the maximum longitude, the minimum longitude, the maximum latitude, and the minimum latitude are enclosed into rectangles, and then the rectangles are divided into several grids, and then the grids are labeled. The value of the number is the grid number of the location of the point.

Algorithm 3: Temporal-Federated learning.

Temperal-Clustering Input:
    Data set $D = \{D_1, D_2, \cdots, D_n\}$
    x, the number of groups to be divided
Temporal-Clustering Output:
    Result $7x$ client groups $G = \{G_1, G_2, \cdots, G_{7x}\}$
    i^{th} group with j clients $G_i = \{D_{i1}, D_{i2}, \cdots, D_{ij}\}$
1: Update the global parameter w to each client group
2: **for** each round $t \in [1, T]$ **do**
3:    randomly choose $G_1, G_2, \cdots, G_m$ from G
4:    **for** each client group G_x **do**
5:      ClientTrain(w, D_x);
6:    **end for**
7:    i^{th} weight for i^{th} client group $\alpha_i = \frac{n_i}{n_{total}}$
8:    $w_t = \frac{\sum_{i=1}^m \alpha_i w_i}{m}$
9: **end for**

TABLE 11.1 Features.

Feature Name	Feature Meaning
number	Trajectory grid number
datetime	Time between now and the first point
lng	Longitude
lat	Latitude
q_lng	Quantile of longitude
q_lat	Quantile of latitude
CVlng	CV of longitude
CVlat	CV of latitude
Δt	Time difference with last point
locC	Distance difference with last point
s	Speed
rot	Rotation rate
f(x)'	f(x)' of the trajectory curve
f(x)''	f(x)'' of the trajectory curve
ρ	Radius of curvature
Ω	Angular velocity

- **datetime:** The time from the start of the trajectory to the current position. $datetime_2$ is the time of this point, $datetime_1$ is the time of the first point of the trajectory.

$$datetime = datetime_2 - datetime_1 \tag{11.10}$$

- **q_lng/q_lat:** The wave rate of the longitude/latitude is

$$wave\_rate = \frac{quantile(x, 0.9) - quantile(x, 0.1) + 1}{quantile(x, 0.1) + 1} \tag{11.11}$$

- **CVlng/CVlat:** Coefficient of variation of trajectory. STD is the standard deviation and Mean is the average value

$$CV = \frac{STD}{Mean} \tag{11.12}$$

- **Δ t:** The time from the last point to the current point. $datetime_2$ is the time of this point, $datetime_1$ is the time of the previous point

$$\Delta t = datetime_i - datetime_{i-1} \tag{11.13}$$

- **locC:** Absolute distance between the point and the previous point:

$$locC = \sqrt{(lat_{xi} - lat_{xi-1})^2 + (lon_{xi} - lon_{xi-1})^2} \tag{11.14}$$

- **f(x)':** The first derivative of the trajectory curve

$$f(x)' = \frac{\Delta y}{\Delta x} \tag{11.15}$$

- **f(x)":** The second derivative of the trajectory curve

$$f(x)'' = \frac{\Delta f(x)'}{\Delta x} \tag{11.16}$$

- **ρ:** Radius of curvature.

$$\rho = \frac{|y''|}{\left(1 + y'^2\right)^{3/2}} \tag{11.17}$$

- **Ω:** Angular velocity.

$$\Omega = \frac{s}{\rho} \tag{11.18}$$

11.8.3 Results

The data set we used is the public data set of the Didi Gaia program. We used data of Chengdu from November 15 to November 21, 2021. The dataset has the feature of traj_id, user_id, longitude, latitude, and datetime. Traj_id and user_id are serial numbers randomly generated by the company, in order to protect user privacy, longitude and latitude are the latitude and longitude of the current point of the track, and datetime is the time when the current point is recorded. We took about 100,000 points from the dataset for a total of 2,797 trajectories. We choose 60% as the training set, 20% as the validation set, and 20% as the test set.

We set some default parameters as follows:

- **num_of_clients:** We set num_of_clients to 6 because we have only six days of data.
- **cfraction:** The proportion of the selected training clients in each communicating round. We set it to 0.4.
- **epoch:** The number of training rounds for each client during local training. We set it to 70.
- **batchsize:** We set it to 40, which we think is appropriate for our data size.
- **learning_rate:** Set to 0.01.
- **num_communicate:** Number of training rounds. In our experiment, training effects of different models at different stages were compared by changing this parameter.

We run our code using PyCharm, an optimizer for optim.SGD and Loss Function for cross_entropy.

We use the proposed federated learning model in combination with a standard transformer for trajectory prediction, predicting the next target point of the trajectory. When measuring the effectiveness of the model, we removed the last point of each track in the test set and predicted the coordinate of the last track point through the previous track point. We mainly evaluated the predicted accuracy as our index. We also carried out and presented other evaluations: top five accuracy, recall score, precision, and f1-score (Fig. 11.12).

By changing communication round, we compared the accuracy of prediction of temporal federated learning, spatial federated learning and ordinary federated learning, as shown in Fig. 11.12. It can be found that, although there is little difference in the effects of the three models in the early stage, when communicating round ≥ 6, temporal and spatial federated learning begins to show better results. Other evaluations are also consistent with this result.

For the combination of spatiotemporal model, we also compare it with the common federated model. Evaluation scores are shown in the Table 11.2.

In this training process, the clients are divided into six parts by time first, and then five parts by space in each group by K-means, forming 30 clients. We set the grid number to 50, the more the total grid number is, the lower score and the higher accuracy that we will achieve. Using the FedWvg method, it can be seen that the seven evaluation indicators of this model are better than those of traditional federated learning.

We made some visual analysis. In the test set, we randomly selected a few trajectories, compared their original terminal point location with the predicted point location, and drew them in the same color. As shown in Fig. 11.13, we box or connect example pairs of points that are successful in trajectory prediction.

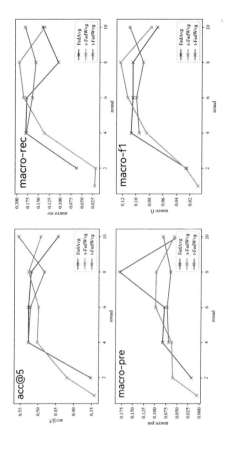

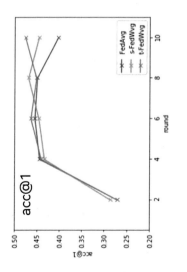

FIGURE 11.12 Model performance.

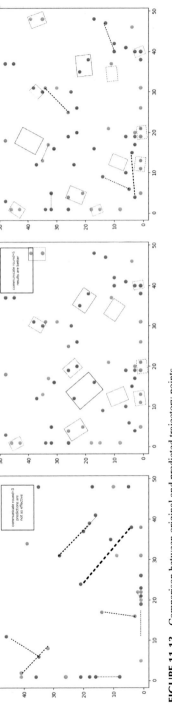

FIGURE 11.13 Comparison between original and predicted trajectory points.

TABLE 11.2 Comparison.

	st_FedWvg	FedAvg
macro-pre	0.0692	0.0528
macro-rec	0.1594	0.1406
macro-f1	0.0945	0.0737
acc@1	0.4598	0.4168
acc@5	0.517	0.4794
acc@10	0.6029	0.5653
acc@20	0.7048	0.687

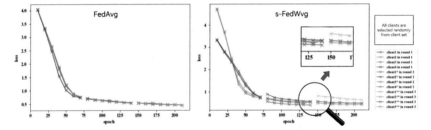

FIGURE 11.14   Loss of FedAvg and s-FedWvg.

In the index evaluation, it is impossible to completely predict the exact location, so we divide the area into several grids, and we consider the prediction as successful as long as it is in the same grid. The smaller each grid area is, the less score but the higher accuracy we will get.

In our model, we can't simply use the model convergence effect as the basis of model training. Take s-FedWvg as an example. Because different clients are randomly selected in each communication round and each group has different characteristics, the performance of loss score for each group in the same round is quite different. As shown in Fig. 11.14, loss scores do not form smooth curves in s-FedWvg because each client has its own characteristics. But, in FedAvg, because data are randomly selected, the curve is smooth and each client in one communication round performs similarly. Therefore, we mainly use the index score as the evaluation of the training model.

11.9 Exercises

- What are the motivations for using federated learning and edge computing? What are their common properties and the common challenges they address?
- What is the difference between horizontal federated learning and vertical federated learning?
- Write down the pseudo-code for FedAvg algorithm.

- In case study 1, we did not write out the AI trigger algorithm. Can you write out its pseudo-code? If so, please do.
- Why deploy AI at the edge? What are the benefits or edge AI?
- Do you consider edge computing and federated learning the future of intelligent transportation systems? Why?
- What kind of role does cloud computing play in edge computing?
- List the techniques you learned that support real-time edge AI at both the system level and algorithm level.
- Describe the parallel intelligence designs on an individual edge device versus across multiple edge devices.
- Give two promising examples in transportation operation and autonomous driving that combine federated learning and edge AI.

Chapter 12

Future directions

In this book, we have examined a variety of data used in multiple applications of big data in transportation. Probe vehicle data (connected vehicle data), other mobility data, socio-demographic data, and economic data can provide profound recommendations concerning possible measures and their impact, e.g. for the creation of transport models, or toll data and truck-related restrictions. Mobile sensing data harvested from smartphones and apps, such as motion and usage data, are a valuable source of information, which can help us analyze which route or mode of transport was taken. A proper analysis of integrated transportation data can also provide valuable assistance in addressing urgent and important issues regarding the mobility of the future, e.g., how can new micro-mobility services be effectively integrated into the existing mobility offering? All of the aforementioned analysis cannot be fulfilled without selecting a proper model.

12.1 Future trends of deep learning technologies for transportation

This book introduces multiple representative deep learning models, such as CNN, RNN, GNN, and GAN, that have been widely used in solving transportation problems. Currently, deep learning is perhaps the most effective AI technology for numerous data-driven applications. However, there are still scientists who point to flaws in deep learning where remedies are not clear.

This section will introduce new trends of deep learning that can complement deep learning's current limitations. Those new trends or solutions may not be mentioned in this book, but have the potential to reshape deep learning. Deep learning is a rapidly growing domain in AI. Due to its challenges about size and diversity of data, various deep learning solutions have been suggested, including introducing reasoning or prior knowledge to deep learning, self-supervised learning, etc. On top of the striking performance achieved by deep neural networks, stable, explainable, generalizable, and safe deep learning models might attract more attentions in the near future. Several new directions of deep learning models based on the authors' knowledge and experience, especially those can be used in the transportation domain, are sketched below:

1. **Stable deep learning**: deep learning models have achieved striking performance when testing data and training data share similar distribution, but can significantly fail otherwise. To eliminate the impact of distribution shifts between training and testing data, it is crucial to build performance-promising

deep models when conventional assumptions of the known heterogeneity of training data (e.g., domain labels) and the approximately equal capacities of different domains do not hold. Thus, it is promising to design models to remove the dependencies between features via learning weights for training samples and help deep models get rid of spurious correlations and, in turn, concentrate more on the true connection between discriminative features and labels.

2. **Unsupervised learning**: If systems can determine their own objectives, do reasoning and problem-solving at a more abstract level, great improvements could be achieved.

3. **Symbol-manipulation and the need for hybrid models**: Integration of deep learning with symbolic systems that excel at inference and abstraction could provide better results.

4. **More insight from cognitive and developmental psychology**: Better understanding the innate machinery in humans minds, gaining common sense knowledge and human understanding of narrative could be valuable for developing learning models.

5. **Explainable deep learning**: Explainability goes one step ahead to explain the internal mechanics of a deep learning system in human terms. It explains what is happening and why every step of the way. Recent advances in AI have enabled algorithms to change and develop quickly, making it even more difficult to interpret what is underlying these models.

6. **Few-shot learning**: The advantage of few-shot learning, which is a subfield of machine learning, is being able to work with a small amount of training data. Few-shot learning algorithms are useful to handle with data shortage and computational costs.

12.2 The future of transportation with AI

The transportation domain is undergoing a major change, and much of the change is being driven by artificial intelligence. We have introduced numerous representative transportation applications that achieved superior performance with the help of deep learning technologies. In the future, we can imagine that more AI technologies will be widely and deeply applied in various transportation scenarios at different levels, such as the single-vehicle level, traffic-flow level, and transportation-system level.

1. **Advanced driver-assistance systems**: Many car manufacturers have long since started to implement semi-autonomous driving features in their vehicles, such as advanced driver-assistance systems (ADAS), to help perform parking procedures, ensure control of the vehicle in bad weather conditions, and avoid collisions. The future ADAS solutions will rely more on AI-powered cameras and sensors designed to identify vehicles, obstacles, pedestrians, or passengers' facial expressions through computer vision, alert the driver, and even trigger autonomous actions. With AI-based ADAS, we

can anticipate that adaptive cruise control, forward-collision warning, automotive night vision, traffic sign recognition, driver monitoring systems, and more advanced functionalities will be implemented in new vehicles. But, the related machine learning-research areas will still be popular.

2. **Self-driving vehicles**: To achieve autonomous driving, AI needs to plan and execute actions without the influence of a human driver. AI is equipped to perform the same functions as a human driver. It has recognition and decision-making abilities, sensory functions, and the ability to model data with deep learning algorithms. Armed with these innovations, the AI-powered vehicle can perform autonomously. The IoT-based sensors generate an enormous amount of data, which is then translated into useful insight with the help of AI algorithms, object character recognition, machine learning, and computer vision techniques. The role of AI in autonomous vehicles is to make a vehicle's sensing, prediction, and control exhibit the characteristics of a human driver. AI enables it to see, hear, think, and make decisions by itself using the data that has been gathered by using the components implanted in the vehicle.

3. **Smart parking**: Few things can be more dreadful than traffic and speeding tickets, bit one of them is the endless search for a parking space in an overcrowded lot or dense city streets. AI can come to our aid and make this "mission impossible" a bit easier with the help of collected parking-lot information via cameras and computer vision. Parking guidance and information systems will provide not only parking availability information but the shortest or proper path to the available slots.

4. **Traffic Management and road monitoring**: Traffic management systems might be seen as something less glamorous than other high-tech transportation solutions such as autonomous vehicles. However, these systems still represent one of the most useful manifestations of artificial intelligence in transportation because they make drivers' lives less stressful, help reduce road accidents, and mitigate pollution. The idea is to deploy a widespread network of sensors and cameras to oversee the traffic flow, monitor road conditions, and identify accidents via computer vision. This allows authorities to intervene promptly in the event of accidents, speed up road repair and maintenance operations, and optimize traffic light switching based on vehicle density. How to design and properly deploy multi-agent reinforcement learning-based large-scale traffic network control model might be the next step for traffic management.

12.3 Book extension and future plan

This book briefly introduces basic machine learning and representative deep learning models. These methods have been widely applied in the transportation research domain and applications, including traffic object detection, road traffic prediction, traffic state classification, among other. With the growth of massive traffic data and the rapid development of deep learning research, we can imagine new technologies including connected autonomous vehicle, autonomous driving, digital twins, will fully reform the transportation system.

To ensure that this textbook continues to provide comprehensive and prompt knowledge of machine learning and deep learning applied in the transportation domain, the authors will revise the context of the book based on the ongoing development of new machine learning technologies and their applications. This textbook will also update the exercises, case studies, and sample source code in each chapter based on the feedback of the book readers, including students it is to be hoped, in future editions. To help teachers use this book and further improve the textbooks' quality, the authors will also provide slides to the public and share the latest information on the textbook website.

Bibliography

Alam, M., Moroni, D., Pieri, G., Tampucci, M., Gomes, M., Fonseca, J., Ferreira, J., Leone, G.R., 2018. Real-time smart parking systems integration in distributed its for smart cities. Journal of Advanced Transportation 2018.

Almeida, T., Lourenço, B., Santos, V., 2020. Road detection based on simultaneous deep learning approaches. Robotics and Autonomous Systems 133, 103605.

Amato, G., Carrara, F., Falchi, F., Gennaro, C., Meghini, C., Vairo, C., 2017. Deep learning for decentralized parking lot occupancy detection. Expert Systems with Applications 72, 327–334.

Amato, G., Carrara, F., Falchi, F., Gennaro, C., Vairo, C., 2016. Car parking occupancy detection using smart camera networks and deep learning. In: 2016 IEEE Symposium on Computers and Communication (ISCC). IEEE, pp. 1212–1217.

Ardestani, S.M., Jin, P.J., Volkmann, O., Gong, J., Zhou, Z., Feeley, C., 2016. 3d accident site reconstruction using unmanned aerial vehicles (uav). Technical report.

Association, A.T., et al., 2017. American trucking trends 2017. Transport Topics.

Atwood, J., Towsley, D., 2016. Diffusion-convolutional neural networks. Advances in Neural Information Processing Systems 29.

Avazpour, I., Pitakrat, T., Grunske, L., Grundy, J., 2014. Dimensions and metrics for evaluating recommendation systems. In: Recommendation Systems in Software Engineering. Springer, pp. 245–273.

Barmpounakis, E.N., Vlahogianni, E.I., Golias, J.C., Babinec, A., 2019. How accurate are small drones for measuring microscopic traffic parameters? Transportation Letters 11 (6), 332–340.

Baroffio, L., Bondi, L., Cesana, M., Redondi, A.E., Tagliasacchi, M., 2015. A visual sensor network for parking lot occupancy detection in smart cities. In: 2015 IEEE 2nd World Forum on Internet of Things (WF-IoT). IEEE, pp. 745–750.

Bayraktar, M.E., Arif, F., Ozen, H., Tuxen, G., 2015. Smart parking-management system for commercial vehicle parking at public rest areas. Journal of Transportation Engineering 141 (5), 04014094.

Bengio, Y., Simard, P., Frasconi, P., 1994. Learning long-term dependencies with gradient descent is difficult. IEEE Transactions on Neural Networks 5 (2), 157–166.

Biçici, S., Zeybek, M., 2021. An approach for the automated extraction of road surface distress from a uav-derived point cloud. Automation in Construction 122, 103475.

Bolourian, N., Soltani, M.M., Albahria, A., Hammad, A., 2017. High level framework for bridge inspection using lidar-equipped uav. In: ISARC. Proceedings of the International Symposium on Automation and Robotics in Construction, vol. 34. IAARC Publications.

Breckon, T.P., Barnes, S.E., Eichner, M.L., Wahren, K., 2009. Autonomous real-time vehicle detection from a medium-level uav. In: Proc. 24th International Conference on Unmanned Air Vehicle Systems. Citeseer, pp. 29–37.

Brehar, R.D., Muresan, M.P., Mariţa, T., Vancea, C.-C., Negru, M., Nedevschi, S., 2021. Pedestrian street-cross action recognition in monocular far infrared sequences. IEEE Access 9, 74302–74324.

Bruna, J., Zaremba, W., Szlam, A., LeCun, Y., 2014. Spectral networks and deep locally connected networks on graphs. In: 2nd International Conference on Learning Representations. ICLR 2014.

Bulan, O., Loce, R.P., Wu, W., Wang, Y.R., Bernal, E.A., Fan, Z., 2013. Video-based real-time on-street parking occupancy detection system. Journal of Electronic Imaging 22 (4), 041109.

Bush, V., et al., 1945. As we may think. The Atlantic Monthly 176 (1), 101–108.

Cai, P., Wang, H., Sun, Y., Liu, M., 2022. Dq-gat: towards safe and efficient autonomous driving with deep q-learning and graph attention networks. IEEE Transactions on Intelligent Transportation Systems.

Cakir, F., He, K., Xia, X., Kulis, B., Sclaroff, S., 2019. Deep metric learning to rank. In: Proceedings of the IEEE Conference on Computer Vision and Pattern Recognition, pp. 1861–1870.

Cao, X., Gao, C., Lan, J., Yuan, Y., Yan, P., 2014. Ego motion guided particle filter for vehicle tracking in airborne videos. Neurocomputing 124, 168–177.

Cao, X., Wu, C., Lan, J., Yan, P., Li, X., 2011. Vehicle detection and motion analysis in low-altitude airborne video under urban environment. IEEE Transactions on Circuits and Systems for Video Technology 21 (10), 1522–1533.

Carletti, V., Greco, A., Saggese, A., Vento, M., 2018. Multi-object tracking by flying cameras based on a forward-backward interaction. IEEE Access 6, 43905–43919.

Chakraborty, P., Sharma, A., Hegde, C., 2018. Freeway traffic incident detection from cameras: a semi-supervised learning approach. In: 2018 21st International Conference on Intelligent Transportation Systems (ITSC). IEEE, pp. 1840–1845.

Chan-Edmiston, S., Fischer, S., Sloan, S., Wong, M., et al., 2020. Intelligent transportation systems (its) joint program office: strategic plan 2020–2025. Technical report, United States. Department of Transportation. Intelligent transportation . . . .

Che, Z., Purushotham, S., Cho, K., Sontag, D., Liu, Y., 2018. Recurrent neural networks for multivariate time series with missing values. Scientific Reports 8 (1), 1–12.

Chen, A.Y., Chiu, Y.-L., Hsieh, M.-H., Lin, P.-W., Angah, O., 2020a. Conflict analytics through the vehicle safety space in mixed traffic flows using uav image sequences. Transportation Research. Part C, Emerging Technologies 119, 102744.

Chen, C., Liu, B., Wan, S., Qiao, P., Pei, Q., 2020b. An edge traffic flow detection scheme based on deep learning in an intelligent transportation system. IEEE Transactions on Intelligent Transportation Systems 22 (3), 1840–1852.

Chen, J., Ding, G., Yang, Y., Han, W., Xu, K., Gao, T., Zhang, Z., Ouyang, W., Cai, H., Chen, Z., 2021. Dual-modality vehicle anomaly detection via bilateral trajectory tracing. In: Proceedings of the IEEE/CVF Conference on Computer Vision and Pattern Recognition, pp. 4016–4025.

Chen, L., Ma, N., Wang, P., Li, J., Wang, P., Pang, G., Shi, X., 2020c. Survey of pedestrian action recognition techniques for autonomous driving. Tsinghua Science and Technology 25 (4), 458–470.

Chen, M., Yu, X., Liu, Y., 2019a. Mpe: a mobility pattern embedding model for predicting next locations. World Wide Web 22 (6), 2901–2920.

Chen, S., Laefer, D.F., Mangina, E., Zolanvari, I., Byrne, J., 2019. Uav bridge inspection through evaluated 3d reconstructions.

Chen, X., Li, Z., Yang, Y., Qi, L., Ke, R., 2020d. High-resolution vehicle trajectory extraction and denoising from aerial videos. IEEE Transactions on Intelligent Transportation Systems 22 (5), 3190–3202.

Chen, X., Wu, S., Shi, C., Huang, Y., Yang, Y., Ke, R., Zhao, J., 2020e. Sensing data supported traffic flow prediction via denoising schemes and ann: a comparison. IEEE Sensors Journal 20 (23), 14317–14328.

Chen, Y., Lv, Y., Wang, F.-Y., 2019c. Traffic flow imputation using parallel data and generative adversarial networks. IEEE Transactions on Intelligent Transportation Systems 21 (4), 1624–1630.

Chen, Y.-y., Lv, Y., Li, Z., Wang, F.-Y., 2016. Long short-term memory model for traffic congestion prediction with online open data. In: Intelligent Transportation Systems (ITSC), 2016 IEEE 19th International Conference on. IEEE, pp. 132–137.

Chen, Z., Zhang, J., Tao, D., 2019d. Progressive lidar adaptation for road detection. IEEE/CAA Journal of Automatica Sinica 6 (3), 693–702.

Cheng, B., Collins, M.D., Zhu, Y., Liu, T., Huang, T.S., Adam, H., Chen, L.-C., 2020a. Panoptic-deeplab: a simple, strong, and fast baseline for bottom-up panoptic segmentation. In: Proceedings of the IEEE/CVF Conference on Computer Vision and Pattern Recognition, pp. 12475–12485.

Cheng, Y., Rau, S., Srivastava, A., Li, S., Parker, S.T., Perry, E., Ahn, S., Noyce, D.A., 2020b. Data archiving and performance measurement for a multi-state truck parking information management system (tpims). In: International Conference on Transportation and Development 2020. American Society of Civil Engineers, Reston, VA, pp. 251–260.

Cho, K., van Merrienboer, B., Gulcehre, C., Bahdanau, D., Bougares, F., Schwenk, H., Bengio, Y., 2014. Learning phrase representations using rnn encoder–decoder for statistical machine translation. In: Proceedings of the 2014 Conference on Empirical Methods in Natural Language Processing (EMNLP), pp. 1724–1734.

Cho, W., Park, S., Kim, M.-j., Han, S., Kim, M., Kim, T., Kim, J., Paik, J., 2018. Robust parking occupancy monitoring system using random forests. In: 2018 International Conference on Electronics, Information, and Communication (ICEIC). IEEE, pp. 1–4.

Constantinescu, S.-G., Nedelcut, F., 2011. Uav systems in support of law enforcement forces. In: International Conference of Scientific Paper AFASES, vol. 2011, pp. 1211–1219.

Cordts, M., Omran, M., Ramos, S., Rehfeld, T., Enzweiler, M., Benenson, R., Franke, U., Roth, S., Schiele, B., 2016. The cityscapes dataset for semantic urban scene understanding. In: Proceedings of the IEEE Conference on Computer Vision and Pattern Recognition, pp. 3213–3223.

Cortes, C.E., Lavanya, R., Oh, J.-S., Jayakrishnan, R., 2002. General-purpose methodology for estimating link travel time with multiple-point detection of traffic. Transportation Research Record 1802 (1), 181–189.

Cui, Z., Ke, R., Pu, Z., Wang, Y., 2020. Stacked bidirectional and unidirectional lstm recurrent neural network for forecasting network-wide traffic state with missing values. Transportation Research. Part C, Emerging Technologies 118, 102674.

Cui, Z., Long, Y., 2019. Perspectives on stability and mobility of transit passenger's travel behaviour through smart card data. IET Intelligent Transport Systems 13 (12), 1761–1769.

Cybenko, G., 1989. Approximation by superpositions of a sigmoidal function. Mathematics of Control, Signals and Systems 2 (4), 303–314.

Daganzo, C.F., 1994. The cell transmission model: a dynamic representation of highway traffic consistent with the hydrodynamic theory. Transportation Research. Part B: Methodological 28 (4), 269–287.

Da'u, A., Salim, N., 2019. Recommendation system based on deep learning methods: a systematic review and new directions. Artificial Intelligence Review, 1–40.

Defferrard, M., Bresson, X., Vandergheynst, P., 2016. Convolutional neural networks on graphs with fast localized spectral filtering. Advances in Neural Information Processing Systems 29.

Djenouri, Y., Zimek, A., Chiarandini, M., 2018. Outlier detection in urban traffic flow distributions. In: 2018 IEEE International Conference on Data Mining (ICDM). IEEE, pp. 935–940.

Dreyfus, S.E., Bellman, R., 1962. Applied Dynamic Programming. Princeton University Press.

Duan, Y., Lv, Y., Wang, F.Y., 2016. Travel time prediction with LSTM neural network. In: IEEE Conference on Intelligent Transportation Systems, Proceedings. ITSC. IEEE, pp. 1053–1058.

Erkent, Ö., Laugier, C., 2020. Semantic segmentation with unsupervised domain adaptation under varying weather conditions for autonomous vehicles. IEEE Robotics and Automation Letters 5 (2), 3580–3587.

Erlik Nowruzi, F., El Ahmar, W.A., Laganiere, R., Ghods, A.H., 2019. In-vehicle occupancy detection with convolutional networks on thermal images. In: 2019 IEEE/CVF Conference on Computer Vision and Pattern Recognition Workshops (CVPRW). Long Beach, CA, USA, pp. 941–948.

Everingham, M., Van Gool, L., Williams, C.K., Winn, J., Zisserman, A., 2010. The Pascal visual object classes (voc) challenge. International Journal of Computer Vision 88 (2), 303–338.

Fan, R., Jiao, J., Pan, J., Huang, H., Shen, S., Liu, M., 2019. Real-time dense stereo embedded in a UAV for road inspection. In: 2019 IEEE/CVF Conference on Computer Vision and Pattern Recognition Workshops (CVPRW). Long Beach, CA, USA, pp. 535–543.

Fan, R., Wang, H., Cai, P., Liu, M., 2020. Sne-roadseg: incorporating surface normal information into semantic segmentation for accurate freespace detection. In: European Conference on Computer Vision. Springer, pp. 340–356.

Fang, X., Huang, J., Wang, F., Zeng, L., Liang, H., Wang, H., 2020. Constgat: contextual spatial-temporal graph attention network for travel time estimation at baidu maps. In: Proceedings of the 26th ACM SIGKDD International Conference on Knowledge Discovery & Data Mining, pp. 2697–2705.

Farag, W., 2020. Real-time detection of road lane-lines for autonomous driving. Recent Advances in Computer Science and Communications (Formerly: Recent Patents on Computer Science) 13 (2), 265–274.

Feng, J., Rong, C., Sun, F., Guo, D., Li, Y., 2020. Pmf: a privacy-preserving human mobility prediction framework via federated learning. Proceedings of the ACM on Interactive, Mobile, Wearable and Ubiquitous Technologies 4 (1), 1–21.

Fernando, T., Denman, S., Sridharan, S., Fookes, C., 2020. Deep inverse reinforcement learning for behavior prediction in autonomous driving: accurate forecasts of vehicle motion. IEEE Signal Processing Magazine 38 (1), 87–96.

Frank, R.J., Davey, N., Hunt, S.P., 2001. Time series prediction and neural networks. Journal of Intelligent & Robotic Systems 31 (1–3), 91–103.

Gal, Y., Ghahramani, Z., 2016. A theoretically grounded application of dropout in recurrent neural networks. In: Advances in Neural Information Processing Systems, pp. 1019–1027.

Geng, B., Tao, D., Xu, C., 2011. Daml: domain adaptation metric learning. IEEE Transactions on Image Processing 20 (10), 2980–2989.

Geng, Y., Cassandras, C.G., 2013. New "smart parking" system based on resource allocation and reservations. IEEE Transactions on Intelligent Transportation Systems 14 (3), 1129–1139.

Gers, F.A., Schmidhuber, J., Cummins, F., 1999. Learning to forget: continual prediction with LSTM.

Glumov, N., Kolomiyetz, E., Sergeyev, V., 1995. Detection of objects on the image using a sliding window mode. Optics and Laser Technology 27 (4), 241–249.

Gomaa, A., Abdelwahab, M.M., Abo-Zahhad, M., 2018. Real-time algorithm for simultaneous vehicle detection and tracking in aerial view videos. In: 2018 IEEE 61st International Midwest Symposium on Circuits and Systems (MWSCAS). IEEE, pp. 222–225.

Goodfellow, I., Bengio, Y., Courville, A., 2016. Deep Learning. MIT Press. http://www.deeplearningbook.org.

Goodfellow, I., Pouget-Abadie, J., Mirza, M., Xu, B., Warde-Farley, D., Ozair, S., Courville, A., Bengio, Y., 2020. Generative adversarial networks. Communications of the ACM 63 (11), 139–144.

Greff, K., Srivastava, R.K., Koutnik, J., Steunebrink, B.R., Schmidhuber, J., 2017. LSTM: a search space odyssey. IEEE Transactions on Neural Networks and Learning Systems 28 (10), 2222–2232.

Grodi, R., Rawat, D.B., Rios-Gutierrez, F., 2016. Smart parking: parking occupancy monitoring and visualization system for smart cities. In: SoutheastCon 2016. IEEE, pp. 1–5.

Haferkamp, M., Al-Askary, M., Dorn, D., Sliwa, B., Habel, L., Schreckenberg, M., Wietfeld, C., 2017. Radio-based traffic flow detection and vehicle classification for future smart cities. In: 2017 IEEE 85th Vehicular Technology Conference (VTC Spring). IEEE, pp. 1–5.

Hammond, D.K., Vandergheynst, P., Gribonval, R., 2011. Wavelets on graphs via spectral graph theory. Applied and Computational Harmonic Analysis 30 (2), 129–150.

Han, H., Ma, W., Zhou, M., Guo, Q., Abusorrah, A., 2020a. A novel semi-supervised learning approach to pedestrian reidentification. IEEE Internet of Things Journal 8 (4), 3042–3052.

Han, H., Zhou, M., Shang, X., Cao, W., Abusorrah, A., 2020b. Kiss+ for rapid and accurate pedestrian re-identification. IEEE Transactions on Intelligent Transportation Systems 22 (1), 394–403.

Haque, K., Mishra, S., Paleti, R., Golias, M.M., Sarker, A.A., Pujats, K., 2017. Truck parking utilization analysis using gps data. Journal of Transportation Engineering. Part A, Systems 143 (9), 04017045.

Hard, A., Rao, K., Mathews, R., Ramaswamy, S., Beaufays, F., Augenstein, S., Eichner, H., Kiddon, C., Ramage, D., 2018. Federated learning for mobile keyboard prediction. ArXiv preprint. arXiv:1811.03604.

He, K., Gkioxari, G., Dollár, P., Girshick, R., 2017. Mask r-cnn. In: Proceedings of the IEEE International Conference on Computer Vision, pp. 2961–2969.

He, S., Luo, H., Chen, W., Zhang, M., Zhang, Y., Wang, F., Li, H., Jiang, W., 2020. Multi-domain learning and identity mining for vehicle re-identification. In: Proceedings of the IEEE/CVF Conference on Computer Vision and Pattern Recognition Workshops, pp. 582–583.

Heinrich, H.W., et al., 1941. Industrial Accident Prevention. A Scientific Approach, second edition.

Henaff, M., Bruna, J., LeCun, Y., 2015. Deep convolutional networks on graph-structured data. ArXiv preprint. arXiv:1506.05163.

Ho, T.-J., Chung, M.-J., 2016. An approach to traffic flow detection improvements of non-contact microwave radar detectors. In: 2016 International Conference on Applied System Innovation (ICASI). IEEE, pp. 1–4.

Hochreiter, S., Schmidhuber, J., 1997. Long short-term memory. Neural Computation 9 (8), 1735–1780.

Hoi, S.C., Liu, W., Chang, S.-F., 2010. Semi-supervised distance metric learning for collaborative image retrieval and clustering. ACM Transactions on Multimedia Computing, Communications, and Applications (TOMM) 6 (3), 1–26.

Hoi, S.C., Liu, W., Lyu, M.R., Ma, W.-Y., 2006. Learning distance metrics with contextual constraints for image retrieval. In: 2006 IEEE Computer Society Conference on Computer Vision and Pattern Recognition (CVPR'06), vol. 2. IEEE, pp. 2072–2078.

Huang, H., Savkin, A.V., Huang, C., 2021. Decentralized autonomous navigation of a uav network for road traffic monitoring. IEEE Transactions on Aerospace and Electronic Systems 57 (4), 2558–2564.

Huang, T.-W., Cai, J., Yang, H., Hsu, H.-M., Hwang, J.-N., 2019. Multi-view vehicle re-identification using temporal attention model and metadata re-ranking. In: CVPR Workshops, vol. 2.

Huang, X., He, P., Rangarajan, A., Ranka, S., 2020. Intelligent intersection: two-stream convolutional networks for real-time near-accident detection in traffic video. ACM Transactions on Spatial Algorithms and Systems (TSAS) 6 (2), 1–28.

Hung, W.-C., Tsai, Y.-H., Liou, Y.-T., Lin, Y.-Y., Yang, M.-H., 2018. Adversarial learning for semi-supervised semantic segmentation. ArXiv preprint .arXiv:1802.07934.

Ibrahim, M.R., Haworth, J., Christie, N., Cheng, T., 2021. Cyclingnet: detecting cycling near misses from video streams in complex urban scenes with deep learning. IET Intelligent Transport Systems 15 (10), 1331–1344.

Ismail, K., Sayed, T., Saunier, N., Lim, C., 2009. Automated analysis of pedestrian–vehicle conflicts using video data. Transportation Research Record 2140 (1), 44–54.

Jayaraman, S.K., Tilbury, D.M., Yang, X.J., Pradhan, A.K., Robert, L.P., 2020. Analysis and prediction of pedestrian crosswalk behavior during automated vehicle interactions. In: 2020 IEEE International Conference on Robotics and Automation (ICRA). IEEE, pp. 6426–6432.

Jeon, Y., Ju, H.-I., Yoon, S., 2018. Design of an lpwan communication module based on secure element for smart parking application. In: 2018 IEEE International Conference on Consumer Electronics (ICCE). IEEE, pp. 1–2.

Jung, S., Song, S., Kim, S., Park, J., Her, J., Roh, K., Myung, H., 2019. Toward autonomous bridge inspection: a framework and experimental results. In: 2019 16th International Conference on Ubiquitous Robots (UR). IEEE, pp. 208–211.

Kaplan, A., Haenlein, M., 2019. Siri, siri, in my hand: who's the fairest in the land? On the interpretations, illustrations, and implications of artificial intelligence. Business Horizons 62 (1), 15–25.

Karaduman, M., Cınar, A., Eren, H., 2019. Uav traffic patrolling via road detection and tracking in anonymous aerial video frames. Journal of Intelligent & Robotic Systems 95 (2), 675–690.

Kataoka, H., Suzuki, T., Oikawa, S., Matsui, Y., Satoh, Y., 2018. Drive video analysis for the detection of traffic near-miss incidents. In: 2018 IEEE International Conference on Robotics and Automation (ICRA). IEEE, pp. 3421–3428.

Kaufmann, S., Kerner, B.S., Rehborn, H., Koller, M., Klenov, S.L., 2018. Aerial observations of moving synchronized flow patterns in over-saturated city traffic. Transportation Research. Part C, Emerging Technologies 86, 393–406.

Ke, R., 2016. A novel framework for real-time traffic flow parameter estimation from aerial videos. PhD thesis.

Ke, R., 2020. Real-Time Video Analytics Empowered by Machine Learning and Edge Computing for Smart Transportation Applications. University of Washington.

Ke, R., Cui, Z., Chen, Y., Zhu, M., Yang, H., Wang, Y., 2020a. Edge computing for real-time near-crash detection for smart transportation applications. ArXiv preprint. arXiv:2008.00549.

Ke, R., Feng, S., Cui, Z., Wang, Y., 2020b. Advanced framework for microscopic and lane-level macroscopic traffic parameters estimation from uav video. IET Intelligent Transport Systems 14 (7), 724–734.

Ke, R., Li, W., Cui, Z., Wang, Y., 2020c. Two-stream multi-channel convolutional neural network for multi-lane traffic speed prediction considering traffic volume impact. Transportation Research Record 2674 (4), 459–470.

Ke, R., Li, Z., Kim, S., Ash, J., Cui, Z., Wang, Y., 2016. Real-time bidirectional traffic flow parameter estimation from aerial videos. IEEE Transactions on Intelligent Transportation Systems 18 (4), 890–901.

Ke, R., Li, Z., Tang, J., Pan, Z., Wang, Y., 2018a. Real-time traffic flow parameter estimation from uav video based on ensemble classifier and optical flow. IEEE Transactions on Intelligent Transportation Systems 20 (1), 54–64.

Ke, R., Liu, C., Yang, H., Sun, W., Wang, Y., 2022. Real-time traffic and road surveillance with parallel edge intelligence. IEEE Journal of Radio Frequency Identification.

Ke, R., Lutin, J., Spears, J., Wang, Y., 2017. A cost-effective framework for automated vehicle-pedestrian near-miss detection through onboard monocular vision. In: Proceedings of the IEEE Conference on Computer Vision and Pattern Recognition Workshops, pp. 25–32.

Ke, R., Zeng, Z., Pu, Z., Wang, Y., 2018b. New framework for automatic identification and quantification of freeway bottlenecks based on wavelet analysis. Journal of Transportation Engineering. Part A, Systems 144 (9), 04018044.

Ke, R., Zhuang, Y., Pu, Z., Wang, Y., 2020d. A smart, efficient, and reliable parking surveillance system with edge artificial intelligence on iot devices. IEEE Transactions on Intelligent Transportation Systems 22 (8), 4962–4974.

Khan, M.A., Ectors, W., Bellemans, T., Janssens, D., Wets, G., 2017. Unmanned aerial vehicle–based traffic analysis: methodological framework for automated multivehicle trajectory extraction. Transportation Research Record 2626 (1), 25–33.

Khan, M.A., Ectors, W., Bellemans, T., Janssens, D., Wets, G., 2018. Unmanned aerial vehicle-based traffic analysis: a case study for shockwave identification and flow parameters estimation at signalized intersections. Remote Sensing 10 (3), 458.

Kipf, T.N., Welling, M., 2017. Semi-supervised classification with graph convolutional networks. arXiv:1609.02907 [abs].

Klauer, C., Dingus, T.A., Neale, V.L., Sudweeks, J.D., Ramsey, D.J., et al., 2006. The impact of driver inattention on near-crash/crash risk: an analysis using the 100-car naturalistic driving study data.

Klein, L.A., Mills, M.K., Gibson, D.R., et al., 2006. Traffic detector handbook: Volume i. Technical report. Turner-Fairbank Highway Research Center.

Krizhevsky, A., Sutskever, I., Hinton, G.E., 2012. Imagenet classification with deep convolutional neural networks. Advances in Neural Information Processing Systems 25, 1097–1105.

Kulis, B., et al., 2012. Metric learning: a survey. Foundations and Trends in Machine Learning 5 (4), 287–364.

Lee, S., Kim, H.G., Ro, Y.M., 2018. Stan: spatio-temporal adversarial networks for abnormal event detection. In: 2018 IEEE International Conference on Acoustics, Speech and Signal Processing (ICASSP). IEEE, pp. 1323–1327.

Lee, S., Park, E., Yi, H., Lee, S.H., 2020. Strdan: synthetic-to-real domain adaptation network for vehicle re-identification. In: Proceedings of the IEEE/CVF Conference on Computer Vision and Pattern Recognition Workshops, pp. 608–609.

Lee, S., Yoon, D., Ghosh, A., 2008. Intelligent parking lot application using wireless sensor networks. In: 2008 International Symposium on Collaborative Technologies and Systems. IEEE, pp. 48–57.

Lei, B., Wang, N., Xu, P., Song, G., 2018. New crack detection method for bridge inspection using uav incorporating image processing. Journal of Aerospace Engineering 31 (5), 04018058.

Leonardi, G., Barrile, V., Palamara, R., Suraci, F., Candela, G., 2018. 3d mapping of pavement distresses using an unmanned aerial vehicle (uav) system. In: International Symposium on New Metropolitan Perspectives. Springer, pp. 164–171.

Li, A., Wang, S., Li, W., Liu, S., Zhang, S., 2020a. Predicting human mobility with federated learning. In: Proceedings of the 28th International Conference on Advances in Geographic Information Systems, pp. 441–444.

Li, J., Cao, X., Guo, D., Xie, J., Chen, H., 2020b. Task scheduling with uav-assisted vehicular cloud for road detection in highway scenario. IEEE Internet of Things Journal 7 (8), 7702–7713.

Li, J., Wei, Y., Liang, X., Zhao, F., Li, J., Xu, T., Feng, J., 2017. Deep attribute-preserving metric learning for natural language object retrieval. In: Proceedings of the 25th ACM International Conference on Multimedia, pp. 181–189.

Li, J., Xu, Z., Fu, L., Zhou, X., Yu, H., 2021a. Domain adaptation from daytime to nighttime: a situation-sensitive vehicle detection and traffic flow parameter estimation framework. Transportation Research. Part C, Emerging Technologies 124, 102946.

Li, J., Ye, D.H., Chung, T., Kolsch, M., Wachs, J., Bouman, C., 2016. Multi-target detection and tracking from a single camera in unmanned aerial vehicles (uavs). In: 2016 IEEE/RSJ International Conference on Intelligent Robots and Systems (IROS). IEEE, pp. 4992–4997.

Li, T., Sanjabi, M., Beirami, A., Smith, V., 2019a. Fair resource allocation in federated learning. ArXiv preprint. arXiv:1905.10497.

Li, Y., Che, P., Liu, C., Wu, D., Du, Y., 2021b. Cross-scene pavement distress detection by a novel transfer learning framework. Computer-Aided Civil and Infrastructure Engineering 36 (11), 1398–1415.

Li, Y., Wu, J., Bai, X., Yang, X., Tan, X., Li, G., Wen, S., Zhang, H., Ding, E., 2020c. Multi-granularity tracking with modularlized components for unsupervised vehicles anomaly detection. In: Proceedings of the IEEE/CVF Conference on Computer Vision and Pattern Recognition Workshops, pp. 586–587.

Li, Y., Yu, R., Shahabi, C., Liu, Y., 2018. Diffusion convolutional recurrent neural network: data-driven traffic forecasting. In: International Conference on Learning Representations (ICLR '18).

Li, Y., Zhu, Z., Kong, D., Xu, M., Zhao, Y., 2019b. Learning heterogeneous spatial-temporal representation for bike-sharing demand prediction. In: Proceedings of the AAAI Conference on Artificial Intelligence, vol. 33, pp. 1004–1011.

Liang, H., Song, H., Li, H., Dai, Z., 2020. Vehicle counting system using deep learning and multi-object tracking methods. Transportation Research Record 2674 (4), 114–128.

Liang, Y., Cui, Z., Tian, Y., Chen, H., Wang, Y., 2018. A deep generative adversarial architecture for network-wide spatial-temporal traffic-state estimation. Transportation Research Record 2672 (45), 87–105.

Lillicrap, T.P., Hunt, J.J., Pritzel, A., Heess, N., Erez, T., Tassa, Y., Silver, D., Wierstra, D., 2015. Continuous control with deep reinforcement learning. ArXiv preprint. arXiv:1509.02971.

Lin, T., Rivano, H., Le Mouël, F., 2017. A survey of smart parking solutions. IEEE Transactions on Intelligent Transportation Systems 18 (12), 3229–3253.

Lin, Y., Saripalli, S., 2012. Road detection from aerial imagery. In: 2012 IEEE International Conference on Robotics and Automation. IEEE, pp. 3588–3593.

Ling, X., Sheng, J., Baiocchi, O., Liu, X., Tolentino, M.E., 2017. Identifying parking spaces & detecting occupancy using vision-based iot devices. In: 2017 Global Internet of Things Summit (GIoTS). IEEE, pp. 1–6.

Liu, B., Adeli, E., Cao, Z., Lee, K.-H., Shenoi, A., Gaidon, A., Niebles, J.C., 2020a. Spatiotemporal relationship reasoning for pedestrian intent prediction. IEEE Robotics and Automation Letters 5 (2), 3485–3492.

Liu, C., Song, Y., Chang, F., Li, S., Ke, R., Wang, Y., 2022. Posture calibration based cross-view & hard-sensitive metric learning for uav-based vehicle re-identification. IEEE Transactions on Intelligent Transportation Systems.

Liu, H., Ma, J., Yan, W., Liu, W., Zhang, X., Li, C., 2018. Traffic flow detection using distributed fiber optic acoustic sensing. IEEE Access 6, 68968–68980.

Liu, J., Li, T., Xie, P., Du, S., Teng, F., Yang, X., 2020b. Urban big data fusion based on deep learning: an overview. Information Fusion 53, 123–133.

Liu, W., Yang, H., Yin, Y., 2014. Expirable parking reservations for managing morning commute with parking space constraints. Transportation Research. Part C, Emerging Technologies 44, 185–201.

Liu, X., Liu, W., Mei, T., Ma, H., 2016. A deep learning-based approach to progressive vehicle re-identification for urban surveillance. In: European Conference on Computer Vision. Springer, pp. 869–884.

Liu, Z., Zhang, W., Gao, X., Meng, H., Tan, X., Zhu, X., Xue, Z., Ye, X., Zhang, H., Wen, S., et al., 2020c. Robust movement-specific vehicle counting at crowded intersections. In: Proceedings of the IEEE/CVF Conference on Computer Vision and Pattern Recognition Workshops, pp. 614–615.

Loewenherz, F., Bahl, V., Wang, Y., 2017. Video analytics towards vision zero. Institute of Transportation Engineers. ITE Journal 87 (3), 25.

Lou, L., Zhang, J., Xiong, Y., Jin, Y., 2019. An improved roadside parking space occupancy detection method based on magnetic sensors and wireless signal strength. Sensors 19 (10), 2348.

Lu, C., Hu, F., Cao, D., Gong, J., Xing, Y., Li, Z., 2019. Virtual-to-real knowledge transfer for driving behavior recognition: framework and a case study. IEEE Transactions on Vehicular Technology 68 (7), 6391–6402.

Luo, S., Zhang, X., Hu, J., Xu, J., 2020. Multiple lane detection via combining complementary structural constraints. IEEE Transactions on Intelligent Transportation Systems 22 (12), 7597–7606.

Luo, Z., Branchaud-Charron, F., Lemaire, C., Konrad, J., Li, S., Mishra, A., Achkar, A., Eichel, J., Jodoin, P.-M., 2018. Mio-tcd: a new benchmark dataset for vehicle classification and localization. IEEE Transactions on Image Processing 27 (10), 5129–5141.

Lyu, N., Wen, J., Duan, Z., Wu, C., 2020. Vehicle trajectory prediction and cut-in collision warning model in a connected vehicle environment. IEEE Transactions on Intelligent Transportation Systems 23 (2), 966–981.

Ma, X., Tao, Z., Wang, Y.Y., Yu, H., Wang, Y.Y., 2015. Long short-term memory neural network for traffic speed prediction using remote microwave sensor data. Transportation Research. Part C, Emerging Technologies 54, 187–197.

Ma, X., Wu, Y.-J., Wang, Y., 2011. Drive net: e-science transportation platform for data sharing, visualization, modeling, and analysis. Transportation Research Record 2215 (1), 37–49.

Makizako, H., Shimada, H., Hotta, R., Doi, T., Tsutsumimoto, K., Nakakubo, S., Makino, K., 2018. Associations of near-miss traffic incidents with attention and executive function among older Japanese drivers. Gerontology 64, 495–502.

Malinovskiy, Y., Saunier, N., Wang, Y., 2012. Pedestrian travel analysis using static bluetooth sensors. Transportation Research Board: Journal of the Transportation Research Board, 1–22.

Mallat, S., 1999. A Wavelet Tour of Signal Processing. Elsevier.

Martchouk, M., Mannering, F., Bullock, D., 2011. Analysis of freeway travel time variability using bluetooth detection. Journal of Transportation Engineering 137 (10), 697–704.

McCord, M.R., Yang, Y., Jiang, Z., Coifman, B., Goel, P.K., 2003. Estimating annual average daily traffic from satellite imagery and air photos: empirical results. Transportation Research Record 1855 (1), 136–142.

McCulloch, W.S., Pitts, W., 1943. A logical calculus of the ideas immanent in nervous activity. The Bulletin of Mathematical Biophysics 5 (4), 115–133.

McMahan, B., Moore, E., Ramage, D., Hampson, S., y Arcas, B.A., 2017. Communication-efficient learning of deep networks from decentralized data. In: Artificial Intelligence and Statistics. PMLR, pp. 1273–1282.

Menouar, H., Guvenc, I., Akkaya, K., Uluagac, A.S., Kadri, A., Tuncer, A., 2017. Uav-enabled intelligent transportation systems for the smart city: applications and challenges. IEEE Communications Magazine 55 (3), 22–28.

Mercader, P., Haddad, J., 2020. Automatic incident detection on freeways based on bluetooth traffic monitoring. Accident Analysis and Prevention 146, 105703.

Mnih, V., Kavukcuoglu, K., Silver, D., Rusu, A.A., Veness, J., Bellemare, M.G., Graves, A., Riedmiller, M., Fidjeland, A.K., Ostrovski, G., et al., 2015. Human-level control through deep reinforcement learning. Nature 518 (7540), 529–533.

Mohan, R., Valada, A., 2021. Efficientps: efficient panoptic segmentation. International Journal of Computer Vision 129 (5), 1551–1579.

Mozaffari, S., Al-Jarrah, O.Y., Dianati, M., Jennings, P., Mouzakitis, A., 2020. Deep learning-based vehicle behavior prediction for autonomous driving applications: a review. IEEE Transactions on Intelligent Transportation Systems 23 (1), 33–47.

Najiya, K., Archana, M., 2018. Uav video processing for traffic surveillence with enhanced vehicle detection. In: 2018 Second International Conference on Inventive Communication and Computational Technologies (ICICCT). IEEE, pp. 662–668.

Newell, G.F., Potts, R.B., 1964. Maintaining a bus schedule. In: Australian Road Research Board (ARRB) Conference, 2nd. 1964, Melbourne, vol. 2(1).

Nieto, R.M., García-Martín, Á., Hauptmann, A.G., Martínez, J.M., 2018. Automatic vacant parking places management system using multicamera vehicle detection. IEEE Transactions on Intelligent Transportation Systems 20 (3), 1069–1080.

Nurullayev, S., Lee, S.-W., 2019. Generalized parking occupancy analysis based on dilated convolutional neural network. Sensors 19 (2), 277.

Oh, J.-S., Jayakrishnan, R., Recker, W., 2002. Section travel time estimation from point detection data.

Ouali, Y., Hudelot, C., Tami, M., 2020. Semi-supervised semantic segmentation with cross-consistency training. In: Proceedings of the IEEE/CVF Conference on Computer Vision and Pattern Recognition, pp. 12674–12684.

Pan, S.J., Yang, Q., 2009. A survey on transfer learning. IEEE Transactions on Knowledge and Data Engineering 22 (10), 1345–1359.

Park, C., Lee, C., Bahng, H., Tae, Y., Jin, S., Kim, K., Ko, S., Choo, J., 2020. St-grat: a novel spatio-temporal graph attention networks for accurately forecasting dynamically changing road speed. In: Proceedings of the 29th ACM International Conference on Information & Knowledge Management, pp. 1215–1224.

Park, W.-J., Kim, B.-S., Seo, D.-E., Kim, D.-S., Lee, K.-H., 2008. Parking space detection using ultrasonic sensor in parking assistance system. In: 2008 IEEE Intelligent Vehicles Symposium. IEEE, pp. 1039–1044.

Pu, Z., Zhu, M., Li, W., Cui, Z., Guo, X., Wang, Y., 2020. Monitoring public transit ridership flow by passively sensing wi-fi and bluetooth mobile devices. IEEE Internet of Things Journal 8 (1), 474–486.

Ren, S., He, K., Girshick, R., Sun, J., 2015. Faster r-cnn: towards real-time object detection with region proposal networks. Advances in Neural Information Processing Systems 28.

Rianto, D., Erwin, I.M., Prakasa, E., Herlan, H., 2018. Parking slot identification using local binary pattern and support vector machine. In: 2018 International Conference on Computer, Control, Informatics and Its Applications (IC3INA). IEEE, pp. 129–133.

Rodrigue, J.-P., 2020. The Geography of Transport Systems. Routledge.

Rodríguez-Canosa, G.R., Thomas, S., Del Cerro, J., Barrientos, A., MacDonald, B., 2012. A real-time method to detect and track moving objects (datmo) from unmanned aerial vehicles (uavs) using a single camera. Remote Sensing 4 (4), 1090–1111.

Rosenblatt, F., 1957. The Perceptron, a Perceiving and Recognizing Automaton Project Para. Cornell Aeronautical Laboratory.

Roshtkhari, M.J., Levine, M.D., 2013. An on-line, real-time learning method for detecting anomalies in videos using spatio-temporal compositions. Computer Vision and Image Understanding 117 (10), 1436–1452.

Rue, H., Held, L., 2005. Gaussian Markov Random Fields: Theory and Applications. Chapman and Hall/CRC.

Rumelhart, D.E., Hinton, G.E., Williams, R.J., 1986. Learning representations by back-propagating errors. Nature 323 (6088), 533–536.

Sadek, B.A., Martin, E.W., Shaheen, S.A., 2020. Forecasting truck parking using Fourier transformations. Journal of Transportation Engineering. Part A, Systems 146 (8), 05020006.

Schuster, M., Paliwal, K.K., 1997. Bidirectional recurrent neural networks. IEEE Transactions on Signal Processing 45 (11), 2673–2681.

Shao, G., Ma, Y., Malekian, R., Yan, X., Li, Z., 2019. A novel cooperative platform design for coupled usv–uav systems. IEEE Transactions on Industrial Informatics 15 (9), 4913–4922.

Shastry, A.C., Schowengerdt, R.A., 2005. Airborne video registration and traffic-flow parameter estimation. IEEE Transactions on Intelligent Transportation Systems 6 (4), 391–405.

Sherstinsky, A., 2020. Fundamentals of recurrent neural network (rnn) and long short-term memory (lstm) network. Physica D. Nonlinear Phenomena 404, 132306.

Shuman, D.I., Narang, S.K., Frossard, P., Ortega, A., Vandergheynst, P., 2013. The emerging field of signal processing on graphs: extending high-dimensional data analysis to networks and other irregular domains. IEEE Signal Processing Magazine 30 (3), 83–98.

Siam, M., Gamal, M., Abdel-Razek, M., Yogamani, S., Jagersand, M., Zhang, H., 2018. A comparative study of real-time semantic segmentation for autonomous driving. In: Proceedings of the IEEE Conference on Computer Vision and Pattern Recognition Workshops, pp. 587–597.

Sifuentes, E., Casas, O., Pallas-Areny, R., 2011. Wireless magnetic sensor node for vehicle detection with optical wake-up. IEEE Sensors Journal 11 (8), 1669–1676.

Simonyan, K., Zisserman, A., 2014. Very deep convolutional networks for large-scale image recognition. ArXiv preprint. arXiv:1409.1556.

Singh, D., Mohan, C.K., 2018. Deep spatio-temporal representation for detection of road accidents using stacked autoencoder. IEEE Transactions on Intelligent Transportation Systems 20 (3), 879–887.

Spears, J., Lutin, J., Wang, Y., Ke, R., Clancy, S.M., 2017. Active safety-collision warning pilot in Washington state. Technical report.

Sprung, M.J., et al., 2018. Freight facts and figures 2017.

Stehly, L., Campillo, M., Shapiro, N., 2007. Travel time measurements from noise correlation: stability and detection of instrumental time-shifts. Geophysical Journal International 171 (1), 223–230.

Sultani, W., Chen, C., Shah, M., 2018. Real-world anomaly detection in surveillance videos. In: Proceedings of the IEEE Conference on Computer Vision and Pattern Recognition, pp. 6479–6488.

Sun, S., Zhang, C., Yu, G., 2006. A bayesian network approach to traffic flow forecasting. IEEE Transactions on Intelligent Transportation Systems 7 (1), 124–132.

Sun, W., Stoop, E., Washburn, S.S., 2018. Evaluation of commercial truck parking detection for rest areas. Transportation Research Record 2672 (9), 141–151.

Sutton, R.S., McAllester, D.A., Singh, S.P., Mansour, Y., 2000. Policy gradient methods for reinforcement learning with function approximation. Advances in Neural Information Processing Systems, 1057–1063.

Taccari, L., Sambo, F., Bravi, L., Salti, S., Sarti, L., Simoncini, M., Lori, A., 2018. Classification of crash and near-crash events from dashcam videos and telematics. In: 2018 21st International Conference on Intelligent Transportation Systems (ITSC). IEEE, pp. 2460–2465.

Talebpour, A., Mahmassani, H.S., Mete, F., Hamdar, S.H., 2014. Near-crash identification in a connected vehicle environment. Transportation Research Record 2424 (1), 20–28.

Tao, A., Sapra, K., Catanzaro, B., 2020. Hierarchical multi-scale attention for semantic segmentation. ArXiv preprint. arXiv:2005.10821.

Teutsch, M., Krüger, W., 2012. Detection, segmentation, and tracking of moving objects in uav videos. In: 2012 IEEE Ninth International Conference on Advanced Video and Signal-Based Surveillance. IEEE, pp. 313–318.

Tian, Y., Zhang, K., Li, J., Lin, X., Yang, B., 2018. Lstm-based traffic flow prediction with missing data. Neurocomputing 318, 297–305.

Tieleman, A., Vinke, A., van Alfen, N., van Dijk, J., Pillen, S., van Engelen, B., 2012. Skeletal muscle involvement in myotonic dystrophy type 2. A comparative muscle ultrasound study. Neuromuscular Disorders 22 (6), 492–499.

Treml, M., Arjona-Medina, J., Unterthiner, T., Durgesh, R., Friedmann, F., Schuberth, P., Mayr, A., Heusel, M., Hofmarcher, M., Widrich, M., et al., 2016. Speeding up semantic segmentation for autonomous driving.

Tsao, P., Ik, T.-U., Chen, G.-W., Peng, W.-C., 2018. Stitching aerial images for vehicle positioning and tracking. In: 2018 IEEE International Conference on Data Mining Workshops (ICDMW). IEEE, pp. 616–623.

Turing, A.M., Haugeland, J., 1950. Computing machinery and intelligence. In: The Turing Test: Verbal Behavior as the Hallmark of Intelligence, pp. 29–56.

Vaswani, A., Shazeer, N., Parmar, N., Uszkoreit, J., Jones, L., Gomez, A.N., Kaiser, Ł., Polosukhin, I., 2017. Attention is all you need. In: Advances in Neural Information Processing Systems, pp. 5998–6008.

Veličković, P., Fedus, W., Hamilton, W.L., Liò, P., Bengio, Y., Hjelm, R.D., 2018a. Deep graph infomax. In: International Conference on Learning Representations.

Veličković, P., Cucurull, G., Casanova, A., Romero, A., Li, P., Bengio, Y., 2018b. Graph attention networks. In: International Conference on Learning Representations.

Vital, F.d.A.A, Ioannou, P., Gupta, A., 2020. Survey on intelligent truck parking: issues and approaches. IEEE Intelligent Transportation Systems Magazine.

Vítek, S., Melničuk, P., 2017. A distributed wireless camera system for the management of parking spaces. Sensors 18 (1), 69.

Wang, C., Mahadevan, S., 2011. Heterogeneous domain adaptation using manifold alignment. In: Twenty-Second International Joint Conference on Artificial Intelligence.

Wang, D., Zhang, J., Cao, W., Li, J., Zheng, Y., 2018. When will you arrive? Estimating travel time based on deep neural networks. In: AAAI, vol. 18, pp. 1–8.

Wang, H., He, W., 2011. A reservation-based smart parking system. In: 2011 IEEE Conference on Computer Communications Workshops (INFOCOM WKSHPS). IEEE, pp. 690–695.

Wang, J., Sun, L., 2021. Reducing bus bunching with asynchronous multi-agent reinforcement learning. ArXiv preprint. arXiv:2105.00376.

Wang, K., Li, F., Chen, C.-M., Hassan, M.M., Long, J., Kumar, N., 2021a. Interpreting adversarial examples and robustness for deep learning-based auto-driving systems. IEEE Transactions on Intelligent Transportation Systems.

Wang, W., Peng, Y., Cao, G., Guo, X., Kwok, N., 2020a. Low-illumination image enhancement for night-time uav pedestrian detection. IEEE Transactions on Industrial Informatics 17 (8), 5208–5217.

Wang, W., Qie, T., Yang, C., Liu, W., Xiang, C., Huang, K., 2021b. An intelligent lane-changing behavior prediction and decision-making strategy for an autonomous vehicle. IEEE Transactions on Industrial Electronics 69 (3), 2927–2937.

Wang, X., Qian, Y., Wang, C., Yang, M., 2020b. Map-enhanced ego-lane detection in the missing feature scenarios. IEEE Access 8, 107958–107968.

Wang, Y., Zhang, W., Henrickson, K., Ke, R., Cui, Z., et al., 2016. Digital roadway interactive visualization and evaluation network applications to wsdot operational data usage. Technical report, Washington (State). Dept. of Transportation.

Watkins, C.J., Dayan, P., 1992. Q-learning. Machine Learning 8 (3), 279–292.

Wei, H., Zheng, G., Gayah, V., Li, Z., 2021. Recent advances in reinforcement learning for traffic signal control: a survey of models and evaluation. ACM SIGKDD Explorations Newsletter 22 (2), 12–18.

Wei, K., Li, J., Ma, C., Ding, M., Wei, S., Wu, F., Chen, G., Ranbaduge, T., 2022. Vertical federated learning: challenges, methodologies and experiments. ArXiv preprint. arXiv:2202.04309.

Williams, R.J., 1992. Simple statistical gradient-following algorithms for connectionist reinforcement learning. Reinforcement Learning, 5–32.

Wrenn, C.A., 2017. Can Autonomous Technology Reduce the Driver Shortage in the Commercial Trucking Industry. PhD thesis, Doctoral dissertation. California Southern University.

Wu, J., Wang, X., Xiao, X., Wang, Y., 2021. Box-level tube tracking and refinement for vehicles anomaly detection. In: Proceedings of the IEEE/CVF Conference on Computer Vision and Pattern Recognition, pp. 4112–4118.

Wu, J., Xu, H., Zheng, Y., Tian, Z., 2018. A novel method of vehicle-pedestrian near-crash identification with roadside lidar data. Accident Analysis and Prevention 121, 238–249.

Wu, Q., Huang, C., Wang, S.-y., Chiu, W.-c., 2007. Robust parking space detection considering interspace correlation. In: 2007 IEEE International Conference on Multimedia and Expo. IEEE, pp. 659–662.

Xu, J., Glicksberg, B.S., Su, C., Walker, P., Bian, J., Wang, F., 2021. Federated learning for healthcare informatics. Journal of Healthcare Informatics Research 5 (1), 1–19.

Yamamoto, S., Kurashima, T., Toda, H., 2020. Identifying near-miss traffic incidents in event recorder data. In: Pacific-Asia Conference on Knowledge Discovery and Data Mining. Springer, pp. 717–728.

Yan, G., Yang, W., Rawat, D.B., Olariu, S., 2011. Smartparking: a secure and intelligent parking system. IEEE Intelligent Transportation Systems Magazine 3 (1), 18–30.

Yang, H., Ke, R., Cui, Z., Wang, Y., Murthy, K., 2022. Toward a real-time smart parking data management and prediction (spdmp) system by attributes representation learning. International Journal of Intelligent Systems 37 (8), 4437–4470.

Yang, S., Ma, W., Pi, X., Qian, S., 2019. A deep learning approach to real-time parking occupancy prediction in transportation networks incorporating multiple spatio-temporal data sources. Transportation Research. Part C, Emerging Technologies 107, 248–265.

Yeh, R., Chen, C., Lim, T.Y., Hasegawa-Johnson, M., Do, M.N., 2016. Semantic image inpainting with perceptual and contextual losses. ArXiv preprint. arXiv:1607.07539.

Yu, H., Li, Z., Zhang, G., Liu, P., Wang, J., 2020. Extracting and predicting taxi hotspots in spatiotemporal dimensions using conditional generative adversarial neural networks. IEEE Transactions on Vehicular Technology 69 (4), 3680–3692.

Yuan, Y., Chen, X., Chen, X., Wang, J., 2019. Segmentation transformer: object-contextual representations for semantic segmentation. ArXiv preprint. arXiv:1909.11065.

Zhang, F., Liu, W., Wang, X., Yang, H., 2020a. Parking sharing problem with spatially distributed parking supplies. Transportation Research. Part C, Emerging Technologies 117, 102676.

Zhang, J., Philip, S.Y., 2019. Broad Learning Through Fusions. Springer.

Zhang, K., He, Z., Zheng, L., Zhao, L., Wu, L., 2021. A generative adversarial network for travel times imputation using trajectory data. Computer-Aided Civil and Infrastructure Engineering 36 (2), 197–212.

Zhang, M., Li, H., Wang, L., Wang, P., Tian, S., Feng, Y., 2020b. Overtaking behavior prediction of rear vehicle via lstm model. In: CICTP 2020, pp. 3575–3586.

Zhang, Z., Li, X., Yuan, H., Yu, F., 2013. A street parking system using wireless sensor networks. International Journal of Distributed Sensor Networks 9 (6), 107975.

Zhang, Z., Tao, M., Yuan, H., 2014. A parking occupancy detection algorithm based on amr sensor. IEEE Sensors Journal 15 (2), 1261–1269.

Zhao, Y., Wu, W., He, Y., Li, Y., Tan, X., Chen, S., 2021. Good practices and a strong baseline for traffic anomaly detection. In: Proceedings of the IEEE/CVF Conference on Computer Vision and Pattern Recognition, pp. 3993–4001.

Zhou, H., Kong, H., Wei, L., Creighton, D., Nahavandi, S., 2014. Efficient road detection and tracking for unmanned aerial vehicle. IEEE Transactions on Intelligent Transportation Systems 16 (1), 297–309.

Zhou, H., Kong, H., Wei, L., Creighton, D., Nahavandi, S., 2016. On detecting road regions in a single uav image. IEEE Transactions on Intelligent Transportation Systems 18 (7), 1713–1722.

Zhu, H., Yu, F., 2015. A vehicle parking detection method based on correlation of magnetic signals. International Journal of Distributed Sensor Networks 11 (7), 361242.

Zhu, J., Sun, K., Jia, S., Li, Q., Hou, X., Lin, W., Liu, B., Qiu, G., 2018. Urban traffic density estimation based on ultrahigh-resolution uav video and deep neural network. IEEE Journal of Selected Topics in Applied Earth Observations and Remote Sensing 11 (12), 4968–4981.

Zhu, J.-Y., Park, T., Isola, P., Efros, A.A., 2017. Unpaired image-to-image translation using cycle-consistent adversarial networks. In: Proceedings of the IEEE International Conference on Computer Vision, pp. 2223–2232.

Zhuang, Y., Ke, R., Wang, Y., 2019. Innovative method for traffic data imputation based on convolutional neural network. IET Intelligent Transport Systems 13 (4), 605–613.

Index

Printed in the United States
by Baker & Taylor Publisher Services